THE TERRY LECTURES
BELIEF IN GOD IN AN AGE OF SCIENCE

Other volumes in the Terry Lecture Series available from
Yale University Press

Belief in God
in an
Age of Science

JOHN POLKINGHORNE

Yale University Press   New Haven & London

First published by Yale University Press in 1998.

Published with assistance from the Louis Stern Memorial Fund.

Printed in the United States of America.

Library of Congress Control Number: 2002113898

ISBN 978-0-300-09949-2 (pbk.)

A catalogue record for this book is available from the British Library.

10 9 8 7 6 5 4

The Dwight Harrington Terry Foundation Lectures on Religion in the Light of Science and Philosophy

The deed of gift declares that "the object of this foundation is not the promotion of scientific investigation and discovery, but rather the assimilation and interpretation of that which has been or shall be hereafter discovered, and its application to human welfare, especially by the building of the truths of science and philosophy into the structure of a broadened and purified religion. The founder believes that such a religion will greatly stimulate intelligent effort for the improvement of human conditions and the advancement of the race in strength and excellence of character. To this end it is desired that a series of lectures be given by men eminent in their respective departments, on ethics, the history of civilization and religion, biblical research, all sciences and branches of knowledge which have an important bearing on the subject, all the great laws of nature, especially of evolution . . . also such interpretations of literature and sociology as are in accord with the spirit of this foundation, to the end that the Christian spirit may be nurtured in the fullest light of the world's knowledge and that mankind may be helped to attain its highest possible welfare and happiness upon this earth." The present work constitutes the latest volume published on this foundation.

To the

Society for
Promoting Christian Knowledge
on its three hundredth anniversary
1698–1998

Contents

Preface

Five principal concerns have characterised activity in the past thirty years across the border between science and theology: a rejection of reductionism, partly based on an appeal to science's increasing recognition of the interconnected and holistic character of much physical process; an understanding of an evolutionary universe as being compatible with a theological doctrine of creatio continua; a revival of a cautiously revised form of natural theology; a methodological comparison of science and theology that exhibits their common concern with the attainment of understanding through the search for motivated belief; and speculations concerning how physical process might be sufficiently open to accommodate the acts of agents, both human and divine. Serious books on science and theology are bound to engage with some or all of these themes. I do not feel that I have much to add to what has been written already on the first two of these topics, but the invitation to give the Terry Lectures provided the opportunity to give further thought to the other three.

Chapter 1 concerns natural theology, presented as an insightful, rather than logically demonstrative, discipline. Underlying its argument is the conviction that theism offers the "best explanation" of the many-levelled character of human encounter with reality. The treatment enlarges my previous discussion of this theme, not least by its greater emphasis on the importance of moral and aesthetic experience and the deep-seated human intuition of hope.

In Chapter 2, I describe the cousinly relationship between scientific and theological method. Though in no conventional sense a Kuhnian, I am persuaded by Thomas Kuhn's insistence on the importance of historical enquiry as the testing ground of philosophical theorising. Epistemological questions are to be settled, not by abstract considerations, but by regarding how it is that we actually gain knowledge. Rather than reworking previous general methodological discussions, I have focused on two critical periods of exploration and discovery, one scientific, the other theological. The first is the investigation into the nature of light that eventually led to quantum theory; the second is the Christological controversies of the first five centuries that eventually led to the Chalcedonian definition. I believe that these particularities bring to light interesting analogies between the ways in which science and theology attain their beliefs.

In Chapter 3, I discuss divine action, a theme that has preoccupied many writers on science and theology, particularly in the past ten years. A critical review is given of the proposals that have been made for locating a "causal joint," or ontological openness within process as science describes it, which could afford room for agents to act. Recent results of Ilya Prigogine in relation to non-integrable (fractal-like) solutions are used to carry further the metaphysical conjecture that chaos theory should be interpreted in an open, rather than a deterministic, way.

Chapter 4 surveys prospects for future dialogue. It is argued that the engagement between scientific and theological thinking will become closer, drawing in a greater involvement of theologians and of workers in the human sciences. The religious setting must broaden beyond the Abrahamic faiths to include all religious traditions. It is suggested that their meeting with science may provide the world faiths with a congenial ground of encounter.

One way of characterising the perceived common ground between science and theology is to say that both make a critical realist evaluation of their encounter with their very different subject matters. This has been a popular theme in the writings of scientists who have a serious interest in theology. My own defence of critical realism in science is based on an analysis of the thirty-year history of the discovery of quarks and gluons, in which I had been a minor participant (see my book *Rochester Roundabout*). Chapter 5 distils many features of the scientific quest for understanding, and it suggests that parallels exist in the theological quest for understanding. The chapter, therefore, constitutes a defence of the critical realist thesis, located at a level of generality midway between the specificities of Chapter 2 and a purely abstract line of argument (if that were possible).

The nature of mathematical truths has long been debated. The sense of discovery, to which the mathematicians testify, encourages the belief that there is a realm of reality in which entities like the Mandelbrot set exist everlastingly. Chapter 6 briefly sketches the metaphysical challenge and opportunity that such a view offers.

The book presents a series of variations on a fundamental theme: if reality is generously and adequately construed, then knowledge will be seen to be one; if rationality is generously and adequately construed, then science and theology will be seen as partners in a common quest for understanding.

Acknowledgements

The first four chapters are based on the Terry Lectures, which I gave at Yale University in October, 1996. I am very grateful for the invitation to give these lectures and for the generous hospitality accorded to my wife and myself on this occasion.

This is the last manuscript typed for me by my secretary, Mrs. Josephine Brown, before my retirement as President of Queens' College, Cambridge. She has given it all the skilful care and attention she also gave its predecessors. Mrs. Brown's unfailing help has been an enormous asset to me as an author, and I express my most sincere thanks. I also thank my wife, Ruth, for her help in correcting the proofs.

CHAPTER ONE

Belief in God in an Age of Science

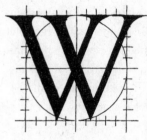HAT does it mean to believe in God today? Different religious communities propose different answers to that fundamental question. I speak from within the Christian tradition, though much of what I say in this chapter would, I believe, find endorsement from my Jewish and Islamic friends. For me, the fundamental content of belief in God is that there is a Mind and a Purpose behind the history of the universe and that the One whose veiled presence is intimated in this way is worthy of worship and the ground of hope. In this chapter, I sketch some of the considerations that persuade me that this is the case.

The world is not full of items stamped "made by God"—the Creator is more subtle than that—but there are two locations where general hints of the divine presence might be expected to be seen most clearly. One is the vast cosmos itself, with its fifteen-billion-year history of evolving development following the big bang. The other is the "thinking reed" of humanity, so insignificant in physical scale but, as Pascal said, superior to all the stars because it alone knows them and itself. The universe and the means by which that universe has become marvellously self-aware—these are the centres of our enquiry.

Those who work in fundamental physics encounter a world whose large-scale structure (as described by cosmology) and small-scale process (as described by quantum theory) are alike characterised by a wonderful order that is expressible in concise and elegant mathematical terms. The distinguished theoretical physicist Paul Dirac, who was not a conventionally religious man, was once asked what was his fundamental belief. He strode to a blackboard and wrote that the laws of nature should be expressed in beautiful equations. It was a fitting affirmation by one whose fundamental discoveries had all come from his dedicated pursuit of mathematical beauty. This use of abstract mathematics as a technique of physical discovery points to a very deep fact about the nature of the universe that we inhabit, and to the remarkable conformity of our human minds to its patterning. We live in a world whose physical fabric is endowed with transparent rational beauty.

Attempts have been made to explain away this fact. No one would deny, of course, that evolutionary necessity will have moulded our ability for thinking in ways that will ensure its adequacy for understanding the world around us, at least to the extent that is demanded by pressures for survival. Yet our surplus intellectual capacity, enabling us to comprehend the microworld of quarks and gluons and the macroworld of big bang cosmology, is on such a scale that it beggars belief that this is simply a fortunate by-product of the struggle for life. Remember that Sherlock Holmes told a shocked Dr. Watson that he didn't care whether the Earth went round the Sun or vice versa, for it had no relevance to the pursuits of his daily life!

Even less plausible, in my view, is the claim sometimes advanced that human beings happen to like mathematical reasoning and so they manipulate their account of physical process into pleasing mathematical shapes.[1] Nature is not so plastic as to be subject to our whim in this way. In 1907, Einstein had what he called "the happiest thought of my life," when he recognised the principle of equivalence, which implied that all entities would move in the same way in a gravitational field. This universality of effect meant that gravity could be expressed as a property of space-time itself; physics could be turned into geometry. Einstein then embarked on a search for a beautiful equation that would determine the relevant geometrical structure. It took him eight years to find it, culminating in the discovery of the theory of general relativity in November 1915. It was a truly beautiful theory but now came the moment of truth. On 18th November, Einstein calculated the prediction made by his theory for the motion of the planet Mercury. He found that it precisely explained a discrepancy in relation to Newton's theory that had baffled astronomers for more than sixty years. Einstein's biographer, Abram Pais, says "This discovery was, I believe, by far the strongest emotional experience in Einstein's scientific life, perhaps in all his life. Nature had spoken to him." Whilst

---

1. A. Pickering, *Constructing Quarks* (Edinburgh University Press, 1984), 413.

the great man himself said, "For a few days, I was beside myself with joyous excitement."[2] It was a great triumph but, if the answer had not come out right, the aesthetic power of the equations of general relativity would have been quite unable in itself to save them from abandonment. It was indeed *nature* that had spoken.

There is no a priori reason why beautiful equations should prove to be the clue to understanding nature; why fundamental physics should be possible; why our minds should have such ready access to the deep structure of the universe. It is a contingent fact that this is true of us and of our world, but it does not seem sufficient simply to regard it as a happy accident. Surely it is a significant insight into the nature of reality. I believe that Dirac and Einstein, in making their great discoveries, were participating in an encounter with the divine. It has become common coinage with contemporary writers about science to invoke, in addressing the general public, the idea of a reading of the Mind of God.[3] It is a small, but significant, sign of the human longing for God that apparently this language helps to sell books. There is much more to the Mind of God than physics will ever disclose, but this usage is not misleading, for I believe that the rational beauty of the cosmos indeed reflects the Mind that holds it in being. The "unreasonable effectiveness of mathematics" in uncovering the structure of the physical world (to use Eugene Wigner's pregnant phrase) is a hint of the presence of the Creator, given to us creatures who are made in the divine image. I do not present this conclusion as a logical demonstration—we are in

2. A. Pais, *Subtle Is the Lord . . .* (Oxford University Press, 1982), 253.
3. See, e. g., P. C. W. Davies, *The Mind of God* (Simon and Schuster, 1992); S. W. Hawking, *A Brief History of Time* (Bantam, 1988).

a realm of metaphysical discourse where such certainty is not available either to believer or to unbeliever—but I do present it as a coherent and intellectually satisfying understanding.

So much for signs of Mind. Where are we to look for signs of Purpose? Before 1859, the answer would have been obvious: in the marvellous adaptation of life to its environment. Charles Darwin, by the publication of *The Origin of Species*, presented us with natural selection as a patient process by which such marvels of "design" could come about, without the intervening purpose of a Designer being at work to bring them into being. At a stroke, one of the most powerful and seemingly convincing arguments for belief in God had been found to be fatally flawed. Darwin had done what Hume and Kant with their philosophical arguments had failed to achieve, abolishing the time-honoured form of the argument from design by exhibiting an apparently adequate alternative explanation.

Since then, two important developments have taken place. One is the realization in the late 1920s that the universe itself has had a history and that notions of evolving complexity apply not only to life on Earth, but to the whole physical cosmos. The other is the acknowledgement that when we take this cosmic history into our reckoning, evolution by itself is not sufficient to account for the fruitfulness of the world. Let me explain.

A convenient slogan-encapsulation of the idea of evolution is to speak of it as resulting from the interplay of chance and necessity. "Chance" stands for the particular contingencies of historical happening. This particular cosmic ripple led to the subsequent condensation of this particular group of galaxies; this particular genetic mutation turned the stream of life in this particular direction rather than another. "Necessity" stands for the lawfully regular environment in which evolution takes place. Without a law of gravity, galaxies would not condense; without reasonably reliable genetic transmission, species would not be established. What we have come to understand is that if this process is to be fruitful on a cosmic scale, then necessity has to take a very specific, carefully prescribed form. Any old world will not do. Most universes that we can imagine would prove boring and sterile in their development, however long their history were to be subjected to the interplay of chance with their specific form of lawful necessity. It is a particular kind of universe which alone is capable of producing systems of the complexity sufficient to sustain conscious life.

This insight, called the Anthropic Principle, has given rise to much discussion.[4] Is it no more than a simple tautology, saying that this universe which contains ourselves must be compatible with our having appeared within its history? For sure that must be so, but it is surprising—and many of us think significant—that this requirement places so tight a constraint on the physical fabric of our world. Although we know by direct experience this universe alone, there are many other possible worlds that we can visit with our scientific imaginations, and almost all of them, we believe, would be infertile. John Leslie, who has given a detailed account of the many processes that depend on the precise character of physical law for their ultimately life-generating effects, has also given a careful

4. J. D. Barrow and F. J. Tipler, *The Anthropic Cosmological Principle* (Oxford University Press, 1986); J. Leslie, *Universes* (Routledge, 1989); see also, J. C. Polkinghorne, *Reason and Reality* (SPCK/Trinity Press International, 1991), chap. 6; *Beyond Science* (Cambridge University Press, 1996), chap. 6.

discussion of what conclusions we might draw from the Anthropic Principle. We are in a realm of discourse where such conclusions depend on the judgement that we have attained a deeper and more comprehensive understanding, rather than that we have deduced a logically unassailable consequence. Leslie believes that it is no more rational to think that no explanation is required of fine anthropic coincidences than it would be to say that my fishing apparatus can accept a fish only exactly 23.2576 inches long and, on casting the rod into the lake, I find that immediately I have a catch, which is simply my good luck—and that's all there is to say about it.[5] The end of the matter for Leslie is: "My argument has been that the fine tuning is evidence, genuine evidence, of the following fact: *that God is real, and/or there are many and varied universes.* And it could be tempting to call the fact an observed one. Observed indirectly, but observed none the less."[6] Either there is one world whose fruitful potential is the expression of divine purpose or there are many worlds, one of which just happens to be right for the evolution of life.

Those who wish to avoid any suggestion of a divine purpose manifested in the fruitful fine tuning of physical law will have to opt for the second of Leslie's alternative explanations.[7] There are a variety of ways in which one might conceive of the existence of such a portfolio of different universes, understood as domains in which different laws of nature are operating. The more plausible accounts will seek to make some appeal to scientific knowledge and will not just rely on the ad

5. Leslie, *Universes*, 9–13.
6. Ibid., 198.
7. A theist could, of course, combine the two options, but personally I find that unappealing.

hoc assumption that there are a lot of separate worlds that just happen to exist.

Many-worlds quantum theory will not do the trick (even if one believed in it, which I do not), for its parallel worlds are simply ones in which quantum events have different specific outcomes and the basic laws of nature are common to them all.[8] Modern ideas about symmetry breaking offer a little more scope. If there is a Grand Unified Theory of the fundamental forces of the universe, then the particular forces that we actually observe, and which are the concern of the Anthropic Principle, will have crystallised out from this highly symmetric ur-state very early in cosmic history, as expansion cooled the world below the relevant transition temperature. The precise details of this symmetry breaking, and the consequent precise force ratios resulting from it, are spontaneously generated through the amplification of tiny random fluctuations. This process need not be literally universal, and the cosmos may be split into vast domains in which different consequences have been realised. The universe observable by us might be a part of one such huge domain, and, of course, in our particular neck of the woods, the force ratios are "by chance" compatible with our evolution. This account is speculative, but motivated, and I am inclined to consider its possibility as far as it goes. That, however, is not very far. One still needs the right sort of Grand Unified Theory for all this to be feasible, and in that respect our universe is still very special compared to the totality of universes that we can imagine.

Moving up on the scale of bold speculation, one might

8. See, for example, J. C. Polkinghorne, *The Quantum World* (Longman/ Princeton University Press, 1984), 67–68; A. Rae, *Quantum Physics: Illusion or Reality?* (Cambridge University Press, 1986), chap. 6.

evoke notions of quantum cosmology which suggest that universes of various kinds are continually appearing as a physical process called inflation blows up microworlds, which have bubbled up as quantum fluctuations in some universal substrate.[9] Proponents of this point of view are sometimes moved to describe our anthropic universe as being "a free lunch." The phrase itself should trigger a cautious evaluation of the offer being made. The cost of this particular cosmic meal is the provision of quantum mechanics itself (a classical Newtonian world would be a perfectly coherent possibility, but a sterile one), and just the right quantum fields to fluctuate in order to produce first inflation and then all the necessary observed forces of nature. This idea is less well established scientifically than the domain option and, in any case, it does not really remove anthropic particularity, for the basic physical laws still have to take certain specific forms which are the necessary foundation of the proposed quantum cosmology.

Beyond this point, speculation becomes rapidly more rash and more desperate. Maybe, the laws of nature themselves fluctuate, so that a vast portfolio of conceivable, or (to us) inconceivable, worlds rise and fall in the relentless exploration of random possibility—occasional patches of transient and varied order in a sea of seething chaos. We have moved far beyond anything that could be called scientific in this exercise of prodigal conjecture. It is time to consider Leslie's other alternative: that there is a divine purpose behind this fruitful universe, whose fifteen-billion-year history has turned a ball of energy into the home of saints and scientists, and that this purpose has been at work in just one world of consistent

9. The quantum vacuum is an active medium owing to fluctuation effects.

physical law (though maybe with domains of different expressions of that law).

Once again the theistic conclusion is not logically coercive, but it can claim serious consideration as an intellectually satisfying understanding of what would otherwise be unintelligible good fortune. It has certainly struck a number of authors in this way, including some who are innocent of any influence from a conventional religious agenda.[10] Such a reading of the physical world as containing rumours of divine purpose, constitutes a new form of natural theology, to which the insight about intelligibility can also be added. This new natural theology differs from the old-style natural theology of Anselm and Aquinas by refraining from talking about "proofs" of God's existence and by being content with the more modest role of offering theistic belief as an insightful account of what is going on. It differs from the old-style natural theology of William Paley and others by basing its arguments not upon particular occurrences (the coming-to-be of the eye or of life itself), but on the character of the physical fabric of the world, which is the necessary ground for the possibility of any occurrence (it appeals to cosmic rationality and the anthropic form of the laws of nature). This shift of focus has two important consequences. The first is that the new-style natural theology in no way seeks to be a rival to scientific explanation but rather it aims to complement that explanation by setting it within a wider and more profound context of understanding. Science rejoices in the rational accessibility of

10. P. C. W. Davies, *God and the New Physics* (Dent, 1983); *Mind of God;* H. Montefiore, *The Probability of God* (SCM Press, 1985); J. C. Polkinghorne, *Science and Creation* (SPCK, 1988), chaps. 1, 2; and n. 4.

the physical world and uses the laws of nature to explain particular occurrences in cosmic and terrestrial history, but it is unable of itself to offer any reason why these laws take the particular (anthropically fruitful) form that they do, or why we can discover them through mathematical insight. The second consequence of this shift from design through making to design built into the rational potentiality of the universe is that it answers a criticism of the old-style natural theology made so trenchantly by David Hume. He had asserted the unsatisfactoriness of treating God's creative activity as the unseen analogue of visible human craft. The new natural theology is invulnerable to this charge of naive anthropomorphism, for the endowment of matter with anthropic potentiality has no human analogy. It is a creative act of a specially divine character.

Physical scientists, conscious of the wonderful order and finely tuned fruitfulness of natural law, have shown significant sympathy with the attitude of the new natural theology. Biological scientists, on the other hand, have been much more reserved. Their attention is focused on the process of the world (particularly, the evolutionary processes of developing terrestrial life) and they pay scant attention to the fundamental physics that underlies that process.[11] They seem to regard it as unproblematic that the chemical raw materials for life are available in our universe. Instead, they look to the variety of life, both in its marvellous fecundity and ingenious strategies for living and also in its wastefulness and suffering, exemplified by the extinction of species and the existence of painful

11. R. Dawkins, *The Blind Watchmaker* (Longman, 1986); *River out of Eden* (Weidenfeld and Nicolson, 1995).

parasitisms. Beneath it all some of them discern no more than the strife of selfish genes struggling for continuing survival. Joy in nature and sorrow at its apparent tragedies are alike, to them, vain human musings on the meaningless tale of cosmic history:

> If the universe were just electrons and selfish genes, meaningless tragedies like the crashing of a bus are exactly what we should expect, along with equally meaningless *good* fortune. Such a universe would be neither evil nor good in intention. It would manifest no intentions of any kind. In a universe of blind physical forces and genetic replication, some people are going to get hurt, other people are going to get lucky, and you won't find any rhyme or reason in it, nor any justice.[12]

Whatever this bleak judgement is, it is clearly not a conclusion of science alone. It was not his knowledge of genetics that enabled Richard Dawkins to make this pronouncement. Rather, it represents his metaphysical judgement on the significance of the scientific story which is presented to us. In fact, it is *science* that is "blind," for as a self-defining methodological strategy it has closed its eyes to the possibility of discerning evil or good or justice or intention. Those who construct metaphysical theories of wider meanings, or lack of meaning, must take science into account, but there is certainly more than one way in which to do so.

The theologian's response to the biologist's unbelief must lie in proposing an alternative interpretation of the history and process of the universe. Here we are concerned, not with metaquestions about the pattern and structure of the physical world, but with metaquestions about how its histori-

12. Dawkins, *River*, 132–33.

cal process is to be understood. This shift of attention corresponds to a transition from natural theology to a theology of nature. We are not now looking to the physical world for hints of God's existence but to God's existence as an aid for understanding why things have developed in the physical world in the manner that they have.

It has been an important emphasis in much recent theological thought about creation to acknowledge that by bringing the world into existence God has self-limited divine power by allowing the other truly to be itself.[13] The gift of Love must be the gift of freedom, the gift of a degree of letting-be, and this can be expected to be true of all creatures to the extent that is appropriate to their proper character. It is in the nature of dense snow fields that they will sometimes slip with the destructive force of an avalanche. It is the nature of lions that they will seek their prey. It is the nature of cells that they will mutate, sometimes producing new forms of life, sometimes grievous disabilities, sometimes cancers. It is the nature of humankind that sometimes people will act with selfless generosity but sometimes with murderous selfishness. That these things are so is not gratuitous or due to divine oversight or indifference. They are the necessary cost of a creation given by its Creator the freedom to be itself. Not all that happens is in accordance with God's will because God has stood back, making metaphysical room for creaturely action. The appar-

---

13. I. G. Barbour, *Religion in an Age of Science* (Harper and Row, 1990), chap. 6; J. Moltmann, *God in Creation* (SCM Press, 1985), chap. 4; A. R. Peacocke, *Creation and the World of Science* (Oxford University Press, 1979), chaps. 2, 3; J. C. Polkinghorne, *Science and Christian Belief* (SPCK, 1994), published simultaneously as *The Faith of a Physicist* (Princeton University Press, 1994), chap. 4; W. H. Vanstone, *Love's Endeavour, Love's Expense* (Darton, Longman and Todd, 1977).

ently ambivalent tale of evolutionary advance and extinction, which Dawkins sees as the sign of a meaningless world of genetic competition, is understood by the Christian as being the inescapably mixed consequence of a world allowed by its Creator to explore and realise, in its own way, its own inherent fruitfulness—to "make itself," to use a phrase as old as the Anglican clergyman Charles Kingsley's response to Darwin's *Origin of Species.* The cruciform pattern of life through death is the way the world is, not only in the familiar tale of biological life on Earth but also cosmically. We are here today because some five billion years ago a star died in the throes of a supernova explosion, scattering into the environment those chemical elements necessary for life, which it had made in the nuclear furnaces of its interior.

The suffering of the world is such that we might be tempted to think that less freedom would be a worthwhile cost to pay for less pain. But do we really wish we had been automata? The well-known free will defence in relation to moral evil asserts that a world with the possibility of sinful people is better than one with perfectly programmed machines. The tale of human evil is such that one cannot make that assertion without a quiver, but I believe that it is true nevertheless. I have added to it the free-process defence, that a world allowed to make itself is better than the puppet theatre of a Cosmic Tyrant.[14] I think that these two defences are opposite sides of the same coin, that our nature is inextricably linked with that of the physical world which has given us birth.

The fact that we wrestle with the problem of pain and

14. J. C. Polkinghorne, *Science and Providence* (SPCK, 1989), chap. 5.

suffering shows us that the cold scientific story of a universe of some losers and some gainers, as presented to us by Dawkins, is far from sufficient to satisfy our human longing to understand and to make sense of the world in which we live. Questions of meaning and justice cannot be removed from the human agenda. The success of the apparently objectified account of science should not tempt us to commit the Enlightenment error of rejecting the subjective as a source of real knowledge. We are *thinking* reeds, and our thoughts far exceed an impersonal evaluation of logical entailment. In fact there seems to be a principle of mutual exclusion between what can be established beyond a peradventure and what is of real significance for the gain of understanding. Kurt Gödel has taught us that even pure mathematics involves an act of intellectual daring, as we commit ourselves to a belief in the unprovable consistency of the axiomatic system under consideration. The Cartesian programme of seeking to found knowledge on the basis of clear and certain ideas has proved to be an unattainable ideal. "Nothing venture, nothing win" is the motto of the intellectual life.

I do not think that this realisation of the necessary precariousness involved in human theorising, condemns us to a post-modernist belief in the personal or communal construction of a variety of views from which we are free to make our à la carte selection. There is a middle way between certainty and relativism, which corresponds to the critical adherence to rationally motivated belief, held with conviction but open to the possibility of correction. Michael Polanyi spoke of such a way when he set out to describe and defend "a frame of mind in which I hold firmly to what I believe to be true, even though

I know that it might conceivably be false."[15] Significantly, he called this epistemological stance "personal knowledge." One of its most striking exemplifications is science itself.

Although science presents its arguments and conclusions in the guise of an objective discourse, its method is, in fact, more subtle and dependent upon acts of personal evaluation.[16] We have already noted that the search for beautiful equations lies at the heart of the success of fundamental physics. The recognition of mathematical beauty resembles other forms of aesthetic experience in that it is hard to describe but, for those endowed with seeing eyes, there is an unmistakable authenticity to it. It involves an acknowledgement of value which must be made by persons and which cannot be reduced to the successful completion of an algorithmic check-list. Yet the long-term fruitfulness of discoveries made in this way, yielding understanding of phenomena far wider than those considered in the original investigation, makes it clear that what is involved here is not the private satisfaction of the aesthetic tastes of a mathematical côterie, but the opening of a window into the reality of the structure of the physical world. Here is the first of a number of signs we must consider which indicate that encounter with value is fundamental to an adequate apprehension of the world in which we live.

It is precisely the recognition of the qualities of elegance, economy and naturalness which solves the problem of the underdetermination of theory by experiment, so often pressed by philosophers of science, who sometimes speak of the process of discovery as if it were a dull routine of fitting

---

15. M. Polanyi, *Personal Knowledge* (Routledge and Kegan Paul, 1958), 214.
16. Cf. J. C. Polkinghorne, *Rochester Roundabout* (Longman/W. H. Freeman, 1989), chap. 21; for what follows, see also Polkinghorne, *Beyond Science*, chaps. 2, 8.

curves to data points. From the point of view of the working scientist, whose thought is consciously or unconsciously controlled by the canons of scientific value, the problem is exactly the reverse—not of selecting from a plethora of possible explanations but of finding one which is adequate to a large swathe of experimental knowledge and which possesses the form of a good scientific theory. Whatever they may write in the formal prose of their published papers, you will find that physicists appeal all the time to value, according belief to an elegant insight long before its experimental verification is completed, and saying of an ugly and contrived idea, "That can't be right." I do not say that such judgements are invariably correct, but they prove to be so to a degree which makes it clear, contrary to the popular presentation, that science is a value-laden activity.

There is another sense in which the community of scientists is one founded on value, and that relates to the honesty and the generosity of intellectual sharing which are the indispensable basis of its activity. Cases of fraud are extremely rare and rightly fatal to the career and reputation of those involved. This is, of course, just a particular professional aspect of general human morality. I believe that it is of the highest significance that we live in a moral world, that we have moral knowledge which tells us that love and truth are better than hatred and lies. I know that much modern criticism is directed to explaining this away as the result of genetic imprinting or tacit communal cultural agreement. There is, no doubt, some truth in these insights, but I cannot think they come anywhere near an adequate account of what is involved. Doubtless parental care for young children has a genetic element of passing on inheritance to future generations, but does that explain a moving case I encountered recently in which a father wished to donate his second kidney to a son, already with his own children, for whom the first transplant had failed? Did Oskar Schindler take great risks to rescue more than a thousand Jews from extermination because of some implicit calculation of genetic advantage? Such a suggestion only shows the desperate poverty of a "morality" of sociobiology. Dawkins himself recognises this to some extent in the closing sentence of *The Selfish Gene:* "We, alone on earth, can rebel against the tyranny of the selfish replicators." [17] I would add, "Not only we can, but we frequently do."

Nor do I think that ethical acts are simply the result of cultural determinations. I know about the selfish nature of the Ik tribe of East Africa, but the very fact that I know about them is due to the atypical character of their morality, perceived as an aberration in the portfolio of human cultures. Of course, there are many variations of detailed practice, but I cannot see the condemnation of the abuse of the young or of the neglect of the elderly as being merely the way we choose to think about these things in our society. Rather, it is a true insight into the way things are, another window into reality. The vulnerable are to be cared for in their vulnerability. All religious traditions value acts of compassion.

The recognition of other forms of value opens further windows into reality. The poverty of an objectivistic account is made only too clear when we consider the mystery of music. From a scientific point of view, it is nothing but vibrations in the air, impinging on the eardrums and stimulating neural currents in the brain. How does it come about that this banal

17. R. Dawkins, *The Selfish Gene* (Oxford University Press, 1976), 215.

sequence of temporal activity has the power to speak to our hearts of an eternal beauty? The whole range of subjective experience, from perceiving a patch of pink, to being enthralled by a performance of the Mass in B Minor, and on to the mystic's encounter with the ineffable reality of the One, all these truly human experiences are at the centre of our encounter with reality and they are not to be dismissed as epiphenomenal froth on the surface of a universe whose true nature is impersonal and lifeless. From the practice of science to the acknowledgement of moral duty, on to aesthetic delight and religious experience, we live in a world which is the carrier of value at all levels of our meeting with it. Only a metaphysical account which is prepared to acknowledge that this is so can be considered to be at all adequate. This is an issue which frequently comes up in conversation with scientific colleagues who are not believers. I am repeatedly seeking to encourage them to take a generous view of the nature of reality, to recognise that a quasi-objective scientific description constitutes a metaphysical net with many holes in it, to reflect in their thinking those same personal qualities that they enjoy and exercise in their lives.

Theism presents an adequately rich basis for understanding the world in that it readily accommodates the many-layered character of a reality shot through with value. Scientific wonder at the rational order of the universe is indeed a partial reading of "the mind of God," as the popular books asserted, speaking better, perhaps, than their authors realised. Yet there is much more to the mind of God than science will ever discover. Our moral intuitions are intimations of the perfect divine will, our aesthetic pleasures a sharing in the Creator's joy, our religious intuitions whispers of God's presence.

The natural understanding of the value-laden character of our world is that there is a supreme Source of Value whose nature is reflected in all that is held in being. Otherwise the pervasive presence of value is hard to understand. I cannot believe that it simply came into being when hominid brains had acquired sufficient complexity to accommodate such thoughts. Rather our ancestors were then able to recognise what had been there from the beginning.

I am presenting here a form of the axiological argument for the existence of God, a twentieth-century version of the fourth way of St. Thomas Aquinas: "Therefore there must also be something which to all beings is the cause of their being, goodness and perfection; and this we call God."[18] The acknowledgement of value is the recognition of worth and our value-laden world testifies to the presence of One who is truly worthy of worship. This is confirmed by our worshipping experience, mediated through public liturgy and private prayer.

The fourth general aspect of contemporary belief in God which I have identified is that there is One who is the ground of hope. At first sight this might seem the most difficult claim to substantiate in an age of science. Transience and death have always been part of the world of human experience. Today, moreover, we realise that mortality characterises the whole universe itself. Not only has it looked very different in the past from its appearance today, but eventually, after many more billions of years, it will change again, ending either in the bang of cosmic collapse or the long-drawn out whimper of an ever-expanding dying world. In my view, the desperate implausi-

18. *Summa Theologiae*, I. 2.3.

bility of Frank Tipler's scenario of "physical eschatology"[19] does nothing to modify the bleakness of this prognosis.

I wish to take with considerable seriousness the implications of this prediction of eventual cosmic futility.[20] In the challenge it presents to belief in God, I do not think it differs greatly from the even more certain assertion of individual human mortality. I have never felt that the perpetuation of the race, or of life itself, or—least of all—of selfish genes, represented sufficient fulfilment to make sense of the history of this world. The fact that we now know that all these carbon-based entities will one day perish, only makes the point more clearly. If cosmic history is no more than the temporary flourishing of remarkable fruitfulness followed by its subsequent decay and disappearance, then I think Macbeth was right and it is indeed a tale told by an idiot.

Yet there is a deep intuition of hope within the human spirit which revolts against such a nihilistic conclusion.[21] The atheist philosopher, Max Horkheimer, expressed a profound longing when he said that the murderer should not triumph over his innocent victim. But if mortality is the final fate of all, then the murderer has a temporary triumph, though he gains but a little of what he has totally denied to his victim. Only God, it seems to me, can take from death the last word. If the human intuition of hope—that all will be well, that the world makes ultimate sense—is not a vain delusion, then God must exist. I would go beyond the Kantian assertion that belief in

19. F. J. Tipler, *The Physics of Immortality* (Macmillan, 1995). For a critique, see Polkinghorne, *Christian Belief/Faith of a Physicist*,164–66.

20. Polkinghorne, *Christian Belief/Faith of a Physicist*, chap. 9.

21. Ibid., 13–14, 18.

God, and in an afterlife, is necessary in order to confirm the moral order of the world, to the claim that the integrity of personal experience itself, based as it is in the significance and value of individual men and women and the ultimate and total intelligibility of the universe, requires that there be an eternal ground of hope who is the giver and preserver of human individuality—the God, as Jesus said, of Abraham, Isaac, and Jacob, "the God, not of the dead, but of the living" (Mk. 12:27)—and the eternally faithful Carer for creation.

Is such a hope a coherent possibility? Here I can only sketch some considerations which I have sought to develop more fully elsewhere.[22] If we regard human beings as psychosomatic unities, as I believe both the Bible and contemporary experience of the intimate connection between mind and brain encourage us to do, then the soul will have to be understood in an Aristotelian sense as the "form," or information-bearing pattern, of the body. Though this pattern is dissolved at death, it seems perfectly rational to believe that it will be remembered by God and reconstituted in a divine act of resurrection. The "matter" of the world to come, which will be the carrier of this reembodiment, will be the transformed matter of the present universe, itself redeemed by God beyond *its* cosmic death. That resurrected universe is not a second attempt by the Creator to produce a world *ex nihilo* but it is the transmutation of the present world in an act of new creation *ex vetere*. God will then truly be "all in all" (1 Cor. 15:28) in a totally sacramental universe whose divine-infused "matter" will be delivered from the transience and decay inherent in present physical process. Such mysterious and exciting be-

22. Ibid., chap. 9.

liefs depend for their motivation not only on the faithfulness of God, but also on Christ's resurrection, understood as the seminal event from which the new creation grows, and indeed also on the detail of the empty tomb, with its implication that the Lord's risen and glorified body is the transmutation of his dead body, just as the world to come will be the transformation of this present mortal world.[23]

Belief in a human destiny beyond death stems not only from the value of individual creatures, but also from the recognised incompleteness of our lives in this world. All of us will die with business unfinished, hurts unhealed, potentialities unrealised. The vision of a continuing process of purification leading to the inexhaustible experience of the vision of the living God, as set out in Dante's *Purgatorio* and *Paradiso*, is a necessary part of the fulfilment which alone makes total sense of the assertion of individual value. I cannot think that mere remembrance, such as process theology's notion of our lives contributing to the filling of the reservoir of divine experience, is an adequate account. It confuses the preservation of the past with the perfection of the future and it gives a diminished description of God's love for Abraham, Isaac, and Jacob, for you and me.

The human paradox is that we perceive so many signs of value and significance conveyed to us in our encounter with reality, yet all meaning is threatened by the apparent finality of death. If the universe is truly a cosmos, if the world is really intelligible through and through, then this life by itself cannot be the whole of the story.

23. Ibid., chap. 6.

I have tried to write of belief in God today as offering us a way of making sense of the broadest possible band of human experience, of uniting in a single account the rich and many-layered encounter that we have with the way things are. I have forsworn false attempts at demonstration and instead I have chosen to rely, as honesty requires, on the persuasiveness of an intellectually satisfying insight. I have suggested that we need to explore with profound seriousness all avenues of our meeting with reality as they open up for us. The impersonal is not to be preferred to the personal, the objective to the subjective, the quantifiable to the symbolic, the repeatable to the unique. All are part of the one world of our experience. I am a passionate believer in the unity of knowledge and I believe that those who are truly seeking an understanding through and through, and who will not settle for a facile and premature conclusion to that search, are seeking God, whether they acknowledge that divine quest or not. Theism is concerned with making *total* sense of the world.[24] The force of its claims depends upon the degree to which belief in God affords the best explanation of the varieties, not just of religious experience, but of all human experience.

The considerations which I have presented in support of that affirmation in this chapter have been of a general character, concerned with our insights of rational beauty, finely-tuned fruitfulness, a value-laden world, and human hopeful defiance in the face of mortality. In the following chapter I seek to continue the defence of Christian belief in a scientific age, by addressing some matters of more specific particularity.

24. Cf. B. Lonergan, *Insight* (Longman, 1957).

CHAPTER TWO

Finding Truth:
Science and Religion Compared

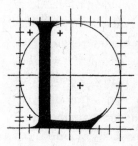ᴇᴛ me tell you a familiar tale from physics: People have wondered about the nature of light for many centuries. In the ancient world, one popular idea was that it was emitted by the eye as humans explored their environment. By the time modern science was established, there were two schools of thought. Isaac Newton and his followers supposed that light might well be composed of a stream of little particles; Christian Huyghens and his followers espoused the idea that it was a form of waves. The question seemed to have been settled in the nineteenth century when the clever experiments of Thomas Young demonstrated the existence of interference effects— the reinforcements and cancellations that result when two trains of waves impinge on each other, either in step (crest augmenting crest) or out of step (trough cancelling crest). James Clerk Maxwell appeared to have spoken the final word when he demonstrated that his theory of electromagnetism

implied the existence of waves travelling with exactly the known velocity of light. There was, of course, the question, waves of what? The answer proposed was the aether, a universal, subtle, and elastic medium that Maxwell, in an encyclopaedia article, described as the best established entity in physics.

The picture changed sharply with the arrival of the twentieth century. It was not the case that nineteenth century insights were completely swept away, but they were found to be only a part of the story, which was richer and more perplexing than it would have been possible for Young and Maxwell to have imagined. Two things happened, in both of which Albert Einstein played a decisive role. First, he used the ideas of Max Planck to show clearly that the way a beam of light ejected electrons from metals could be understood only if it were behaving as if composed of particles of light, to which somewhat later the name of photon was given. Second, by the discovery of special relativity, Einstein revised our notions of the nature of space and time and abolished the possibility of postulating an aether. Both these fundamental discoveries were made in the same year, 1905, while Einstein was employed as a third-class clerk in the Patent Office in Berne!

A period of great confusion followed. The splendid edifice of classical physics, of which Newton and Maxwell had been the chief architects, began to fall apart. Some useful temporary shelters were erected in the ruins by people like Niels Bohr, whose inspired tinkering with classical physics added quantum patches to cover gaping Newtonian holes. This afforded valuable insight but no permanently satisfactory solution. Meanwhile many physicists simply averted their eyes from the distressful paradox of an entity that was sometimes a wave and sometimes a particle.

In 1925, a new revelation came. Almost simultaneously, Erwin Schrödinger and Werner Heisenberg found two very different formulations of what in the end proved to be the same theory. Modern quantum mechanics had come to birth. Its character was further elucidated by the work of Max Born and Paul Dirac. The latter, in 1927, gave a quantum mechanical account of light, producing thereby the first quantum field theory. This discovery resolved the wave/particle paradox in a most satisfactory way, for a quantum field exhibits both discrete quantised properties (particle-like behaviour) and also spread-out field properties (wave-like behaviour).

Although the formalism of quantum theory was well established by the end of the 1920s, its character and consequences have remained subjects of investigation ever since and we have not yet attained a full understanding of all that is implied by it. A question of continuing concern is how the fitful quantum world relates to the reliable world of everyday, in which our measurements are actually made. Quantum theory itself only assigns probabilities for a range of possible outcomes of observation. How does it come about that on a particular occasion of measurement a particular one of these possibilities is actually recorded? Bohr, who played the role of philosopher to the new quantum theory, essentially finessed the problem. The so-called Copenhagen interpretation says little more than that it must happen. Discontent with this summary dismissal of the issue has grown, and the past forty years have seen many competing suggestions of alternative proposals. It is not my present purpose to summarise these, beyond saying that none is wholly satisfactory and the matter remains contentious and unsettled.[1]

Continuing engagement with quantum theory has revealed striking consequences unsuspected at the time of its inception. In 1935, Einstein, with two young collaborators, Boris Podolsky and Nathan Rosen, pointed out that quantum theory appeared to imply a counter-intuitive "togetherness-in-separation" (non-locality) by which two quantum entities that had interacted with each other retained a degree of instantaneous mutual influence, however far they had subsequently separated. Einstein (who was ideologically opposed to modern quantum theory) thought that this was so crazy that it showed the theory was in some way incomplete. Others were not so sure and in the 1980s Alain Aspect and his collaborators were able to show experimentally that this behaviour is indeed present in nature.

The combination of quantum mechanics with special relativity, achieved in the 1920s, has subsequently been found to have properties unsuspected from a knowledge of either theory on its own. The most celebrated of these is the existence of antimatter, suggested by Dirac in 1931. Later that decade, Wolfgang Pauli showed that there is a necessary relationship between spin (the intrinsic angular momentum of particles) and statistics (how they behave in aggregate). In the early 1950s, through the work of Pauli and others, it was realised that the relationship of matter to antimatter is governed by a technical, but very important, principle called the CPT theorem.

1. See J. C. Polkinghorne, *The Quantum World* (Longman/Princeton University Press, 1984), chap. 6; A. Rae, *Quantum Physics: Illusion or Reality?* (Cambridge University Press, 1986).

This long and tangled tale of physics exhibits features found in other quests for understanding. My summary of its characteristics would be:

(1) Moments of radical revision in which new phenomena lead to new insights, so placing the ideas of the past in a novel intellectual setting, transcending previous understanding but still retaining elements in continuity with it. There is a change in the questions to be faced without total abandonment of the answers obtained before. (The transition from Maxwell's understanding of the nature of light to Einstein's, with the consequent recognition of wave/particle duality.)

(2) Periods of confusion in which old and new ideas stand side by side in unresolved tension with each other, and people hold on to experience without being able totally to reconcile the different aspects of it. (Quantum theory from 1900 to 1925.)

(3) Moments of new synthesis and understanding, in which a theory is revealed capable of satisfactorily explaining the new phenomena in a convincing and comprehensive way while, at the same time, treating the old phenomena as particular limiting cases. (The discovery of modern quantum theory.)

(4) A continuing wrestling with unresolved problems, essential for a total understanding of the new theory, but for the moment not capable of final settlement. (The measurement problem in quantum theory.)

(5) Realisations that the new theory has deep implications of a kind unanticipated when it was first conceived (antimatter, non-locality, etc.).

I would add to these characteristics a further comment of a more philosophical, and so contentious, kind. It is the belief of the physicists involved (though not of many of the philosophers who attempt second-order comment on these activities) that what is going on in this process is not merely the extension of a scientific manner of speaking in order to achieve a more empirically adequate account embracing the new phenomena that have been discovered, but the actual uncovering of a more accurate (verisimilitudinous) account of the nature of the physical world. It is the desire for ontological knowledge, and not for mere functional success, which motivates the labour of scientists.

In theology, this kind of story would be called the development of doctrine. Consider these three passages:

> The gospel concerning [God's] Son, who was descended from David according to the flesh and designated Son of God in power according to the Spirit of holiness by his resurrection from the dead, Jesus Christ our Lord (Rom. 1:3-4).

> In the beginning was the Word, and the Word was with God and the Word was God . . . And the Word became flesh and dwelt among us (Jn. 1: 1,14).

> One and the same Son, our Lord Jesus Christ, at once complete in Godhead and complete in manhood, truly God and truly man, . . . one and the same Christ, Son, Lord, Only-begotten, recognised in two natures, without confusion, without change, without division, without separation.

The first is from the epistle to the Romans, where scholars believe Paul is quoting an early Christian formula originating from a period only a few years after the crucifixion. In the second, we have an extract from the well-known prologue to John's gospel, the mature reflection of a profound Christian thinker writing towards the end of the first century. The third is part of the definition issued by the Council of Chalcedon in 451, bringing into play the full resources of Greek philosophy in an effort to articulate orthodox understanding of the status of Christ.

There are clearly immense contrasts between these three statements. The first is a raw response to the resurrection of Jesus. (I have elsewhere given my reasons for believing the resurrection to be an event transcending history but located within it, and I shall not recapitulate those arguments here.)[2] The impact of this vindication of Christ is such that Jesus is spoken of as Son of God, a title of great honour but not one, in Hebrew thought, that carried a necessary connotation of divinity, for it was applied in ancient Israel to the king. The second passage uses imagery of God's active Word, drawn from the Old Testament, evoking also Hebrew ideas of God's Wisdom, and combining these concepts with the Greek notion of the *Logos*, the principle of the rational order of the world. The text constitutes a subtle but clear affirmation, both of the Word's pre-existent divine status, and also of the Word's enfleshing in the life of the earthly Jesus.[3] The third passage uses language, such as that of the two natures, which in its philosophical character is foreign to the style of the New Testament, in the effort to maintain with equal force and clarity both the divine and the human characters of Christ.[4] Here the language of Godhead is used without nuance or qualification.

2. J. C. Polkinghorne, *Science and Christian Belief* (SPCK, 1994), published simultaneously as *The Faith of a Physicist* (Princeton University Press, 1994), chap. 6.

3. Cf. R. E. Brown, *An Introduction to New Testament Christology* (Geoffrey Chapman, 1984), 187–88.

4. For an account of this philosophical struggle, in its character both subtle and confused, see C. Stead, *Philosophy in Christian Antiquity* (Cambridge University Press, 1994), chaps. 16, 17.

It is clear that there is a gradation within this sequence of passages, both in relation to the strength of the claims to status being made and also with respect to the sophistication with which they are expressed. The first passage is an affirmation of faith in God's act in raising Christ and in the one so resurrected; the second a deep reflection on what that affirmation might imply; the third an attempt to articulate belief in the God-man Jesus Christ with all possible metaphysical precision. A central Christological question is whether one does indeed see here a continuous exploration and development of doctrine, or whether one sees instead doctrinal distortion or transmutation, under the pressures of Greek philosophical questioning in the first few centuries. I take the developmental point of view, and I defend that judgement by drawing on analogies that I detect with the physical story with which I began this chapter.[5] I shall organise my argument around the five characteristics I drew from the tale of the developing understanding of the nature of light and the discovery of quantum mechanics.

MOMENTS OF RADICAL REVISION IN WHICH NEW PHENOMENA LEAD TO NEW INSIGHTS

The events which gave rise to Christological struggle and enquiry were the death and resurrection of Jesus. Martin Hengel identifies the twin assertions of the earliest Christian tradition as being "Christ died for us" and "God raised Jesus from the dead."[6] Already in his lifetime, Jesus through

5. See C. F. D. Moule, *The Origin of Christology* (Cambridge University Press, 1977). For my own account of Christology, see *Christian Belief/Faith of a Physicist*, chap. 7.

6. M. Hengel, *Atonement* (SCM Press, 1981).

his words and deeds had put into people's minds the question of who exactly he was. The gospel incident at Caesarea Philippi (Mk. 8:27–38 and parallels) portrays the disciples as reporting popular opinion to the effect that he was a powerful figure returned from the past, Elijah or John the Baptist or one of the prophets. In reply to Jesus' questioning, Peter blurts out the conviction that he is God's anointed and chosen one, the Christ or Messiah. Jesus accepts the designation but binds the disciples to silence. The reason is clear enough, for Jesus speaks immediately of the necessity for suffering, in contradiction to the easy triumph to be expected of a restored Davidic king. Peter can neither understand this nor accept it.

What Jesus had foretold and Peter denied becomes all too plainly true in the rejected figure hanging on the cross in the darkness of Calvary. To all outward appearance it is a scene of complete defeat. Yet the last word lies with God and it is the unanimous conviction of the writers of the New Testament that God uttered that word by raising Jesus from the dead. One of the strong lines of argument for the truth of the resurrection is the astonishing transformation of the disciples from the demoralised defeated men of Good Friday to the confident proclaimers of the Lordship of Christ at Pentecost and beyond, even to the point of martyrdom. *Something* happened to bring that about. I believe it was the resurrection and that if Jesus had not been raised it is probable that we would never have heard of him.

Involved here is the mixture of the continuity and discontinuity of account always present at great turning points of understanding. (Because I discern, both scientifically as well as theologically, the presence of a degree of continuity within the new act of revelation, I do not choose to use Thomas

Kuhn's discontinuous language of paradigm shift.)[7] Old Testament categories, such as Messiah and Suffering Servant and Prophet, are brought into play, but they are given new meanings and presented in unprecedented combinations. I believe that Jesus was not as unreflective about himself as many New Testament theologians seem to suppose, and that he participated in this creative exploration by designating himself the Son of Man, influenced by the apocalyptic figure of Daniel 7, but carrying new overtones of meaning through association with the coming passion and vindication, which I am sure Jesus foresaw in an insightful and trusting way, though not as a detailed glimpse of future history afforded to him.[8]

Just as one can summarise the dilemma about the nature of light at the turn of the century in the single phrase "wave and particle," so one can summarise the Christological dilemma for Jesus' first followers in the phrase "a crucified Messiah." It is the oxymoronic combination of defeat and victory, resulting from the fact that God's way of manifesting saving action is not through naked power but by the acceptance of suffering and the transcendence of death. Crucifixion was not only a painful end. For a Jew it was a sign of God's rejection, since Deuteronomy proclaimed a curse on the one hung on a tree. For a Greek it was the shameful punishment suitable only for slaves and felons. The resurrection showed that all this had to be reassessed, but it is scarcely surprising that St. Paul wrote to the Corinthians that "we preach Christ crucified, a stumbling block to Jews and folly to Greeks." His

---

7. T. Kuhn, *The Structure of Scientific Revolutions*, 2d ed. (University of Chicago Press, 1970).

8. Cf. Brown, *Christology*, 89–100; see also Polkinghorne, *Christian Belief/ Faith of a Physicist*, 98–100.

encounter with the risen Lord enabled him to continue, "but to those who are called, both Jews and Greeks, the power of God and the wisdom of God" (1 Cor. 1:23, 24).

A PERIOD OF CONFUSION IN WHICH OLD AND NEW IDEAS STAND SIDE BY SIDE IN UNRESOLVED TENSION

The writers of the New Testament are struggling to make sense of the aftermath of the events of the crucifixion and resurrection. They believe that Christ died for them and they experience in their lives something of the power of his risen life, which leads them to use metaphors of a new creation or of life from the dead. They are almost all Jews and they know that the God who acted in Jesus is the "one Lord" who is the God of Israel. The New Testament writers are instinctive monotheists. Jesus was a man—some of the writers perhaps knew him, more knew those who knew him—but their experience of his transforming power is such that it is hard to contain talk about Jesus within human categories alone. Very seldom indeed do they dare to say out and out that he is in some sense "God," but divine attributes keep getting attached to him nevertheless.[9] It seems that purely human language just will not do the job that is required.

Let me give one example of what is happening. St. Paul begins many of his letters with a characteristic greeting: "Grace to you and peace from God our Father and the Lord Jesus Christ" (Rom. 1:7, etc.). Something odd is going on here. First of all, God and Jesus are bracketed together in a way which seems most inappropriate for a mere human being.

9. Brown, *Christology*, 171-95.

One could not for a moment imagine Paul, as a pious Jew, substituting his own name to be the one to go alongside that of the divine Father. Second, Jesus is accorded the title "Lord," as he is more than two hundred times in the Pauline writings alone. While the Greek *Kyrios* could be used as a respectful form of address, particularly in the vocative, something more significant than that is being implied by this particular usage. Since YHWH, the divine name, could not be pronounced aloud by a Jew without committing blasphemy, the word "Lord" was substituted for it in the reading of the Hebrew scriptures, as it is in many English translations of the Old Testament today. One begins to see that this title ascribed to Jesus carries with it hints and possibilities going beyond a merely human account of his status. Remember that it is monotheistic Jews who are engaged in this cautious exploration of how one can fittingly speak of Christ.

It is clear that the formula "God our Father and the Lord Jesus Christ" is intellectually unstable. It acknowledges that adequate discourse about Christ must make use of divine language, but it leaves this in unresolved tension with the fundamental Jewish insight of the Oneness and Lordship of God. Much more theological work remains to be done, in the course of which an understanding must be sought of how the Lordship of the Father and the Lordship of Christ are related to each other. What Paul has offered us is a Christological patch on an Old Testament structure that must eventually be rebuilt to accommodate the new insight of the cross and resurrection as God's great saving act. To change the metaphor to one with gospel endorsement, new wine cannot for long be retained in old bottles, otherwise it will soon burst its bonds.

There is an obvious analogy between the Pauline formula

and something like the Bohr theory of the hydrogen atom. The latter was a quantum patch on classical physics, highly instructive as a heuristic device for exploring the new physics, but containing too many internal contradictions to be more than a staging post on the way to a deeper understanding.

It does not trouble me that one cannot find articulated in the New Testament the developed Christian doctrines of the Trinity and the Incarnation. Instead one finds accounts of the foundational events and experiences that set the Church out on the road to the discovery of such doctrines, together with those brilliant provisional insights, such as the Pauline Lordship formula, which are the initial engagement with, but not the completion of, a seeking to come to terms with the new knowledge stemming from the new phenomena of Christ.

MOMENTS OF NEW SYNTHESIS AND UNDERSTANDING IN WHICH A THEORY IS REVEALED

While there is an analogy to be drawn here, it would not be possible to claim that theology has proved as successful in this aspect as it has been the good fortune of science to be. The reasons for the discrepancy are clear enough. Science investigates a physical world that is open to the manipulative investigation of our experimental enquiry. Theology seeks to speak of the God who is to be encountered in awe and obedience, and who is not available to be subjected to our testing interrogation. Science has had to hand the language of mathematics, which we have seen has proved to be perfectly fitted to the description of the fundamental structure of the physical world. Theology has no words adequate to encompass the mystery of the divine nature. It has sought to make use of the open and living language of symbol, but its discourse will never be able fully to encompass God within the limitations of finite human understanding.

Nevertheless, theology has made some attempts at theory-making. The Trinitarian deliberations of the Councils of Nicaea and Constantinople and the Christological deliberations of the Council of Chalcedon are best understood as essays in this mode, attempts to tease out and codify the implications of the unsystematic records of experience and preliminary interpretation which we have seen are found in the pages of the New Testament. These New Testament records are as indispensable to Christian theology as are experimental notebooks to science, but by themselves they no more constitute a theory than do the raw observations and preliminary hunches of a laboratory worker at the bench. I believe that it is extremely important to recognise this character of patristic theology as being a considered reflection on the experience of Christ and his Church and not a project of unlimited and unearthed metaphysical speculation.

In fact, in so far as the Councils were at all able to succeed in their task of theological theory-making, this was mostly in the area of delimiting the range of discussion that could bear acceptable relationship to actual Christian experience. Not to see that the One God was nevertheless known as Father, Son, and Holy Spirit would not, they believed, have been true to the New Testament witness and to the experience of succeeding centuries. Not to recognise in Jesus both the truly human and the truly divine would not have been true to the Christian encounter with Christ. This left the Fathers operating in deeply mysterious territory, but there was no other place in which they felt it was possible for them to take their Christian stand.

Positive affirmations were much harder to come by, as the Chalcedonian definition illustrates, with its emphasis more clearly on where error lies (denying Godhead or manhood, confusing the natures, etc.) than on a positive articulation of what is the case (a reconciling account of the person of the God-man). The best that could be done—then and since— seemed to be a range of partial models affording limited insight, such as Augustine's subtle analogy of the Trinity with the multiple aspects of the deep human psyche or Origen's much more naive use of the iron and the fire as a model of simultaneous human and divine presence. Science has often also found itself in the position of using partial models to gain limited insight, but the comparatively manageable character of its endeavour has often allowed it to extricate itself from this situation and to go on to discover a theory, a candidate for the verisimilitudinous description of the ways things are. Theology has proved to be too difficult for that.

A CONTINUING WRESTLING
WITH UNSOLVED PROBLEMS

This is the normal mode of activity in theological enquiry. Chalcedon certainly did not solve the Christological problem and struggle with its difficulties has continued down the centuries. Some of this activity has been contained within the limits set out by the Chalcedonian Fathers and its focus has been on how it is possible to conceive of the presence in one person of the infinite divine nature and the finite human nature. One solution, particularly pursued by some Anglicans from Charles Gore onwards, has been to speak of a kenosis, a self-emptying of infinite divine qualities by the Word in the acceptance of the finite limits of the enfleshed life of Jesus during the period of his earthly existence.

Others have wished to break the Chalcedonian bounds and to speak of Jesus in human terms as a man inspired by God, and in obedient union with God, to an unparalleled degree. The difference between him and us then becomes a question of intensity rather than ontological distinction. This leads to a kind of functional Christology in which the purpose of Jesus is on the one hand to allow the divine love and power so to transfigure his life that God's nature is made visible to us ("a window into God") and on the other hand so to show us the possibilities of a human life truly lived in communion with God that we are led to seek to share in this experience.[10] This is a view that has appealed to some scientist-theologians, often phrased in evolutionary language. Jesus is described as "the new emergent," the pioneer of the latest development in the upward unfolding of human possibility. Ian Barbour writes in this way and to some extent so does Arthur Peacocke.[11]

There are echoes here of some of the language used in the epistle to the Hebrews (for example, Heb. 2:10, but note also 1:3), but I do not find this way of thinking to be adequate to the witness of the New Testament as a whole. I believe that the insights of a functional Christology ask some of the

10. Cf. J. A. T. Robinson, *The Human Face of God* (SCM Press, 1972).

11. I. G. Barbour, *Religion in an Age of Science* (SCM Press, 1990), 209-14; A. R. Peacocke, *Theology for a Scientific Age*, enlarged ed. (SCM Press, 1993), chaps. 13, 14; for a critique, see J. C. Polkinghorne, *Scientists as Theologians* (SPCK, 1996), chap. 6.

relevant questions but they do not provide the right answers. The work of Christ (what he achieves) is certainly the clue to the nature of Christ. A scientist framing a theory has first to decide what are the phenomena that must be included and explained. The same is true for Christology. Its agenda is set by the functions that Christ fulfils, but a list of these functions is not itself a Christology.

The Church believes that Jesus does not only show us the love of God and encourage us to commit ourselves to the search for the Kingdom but he also died for our sins, bringing us the forgiveness that releases us from the entail of past alienation from God, and through his resurrection he offers us the power of a new life in which to enter into our heritage as children of God. "If we have been united with him in a death like his, we shall certainly be united with him in a resurrection like his" (Rom. 6:5). Much more is involved here than illustration, example, and encouragement, and much more is needed if we are truly to be delivered from the plight of our human condition. Signs of God's love and illustrations of godly living need to have added to them the transforming power of divine life that will enable us to become what God wills for us to be. If Jesus were just the new emergent, how would that help us who have so obviously failed to emerge? But if Jesus is the God-man, the meeting point of divine life and human experience, then there is indeed available to us "a new and living way" (Heb. 10:20). The functions that Jesus fulfils require a corresponding ontological status to make that possible, going beyond the simply human.

I believe that Christian experience of redemption and new life in Christ sets an agenda that requires us to continue to wrestle with the deeply mysterious problems of an orthodox understanding of the incarnation, pointing to an ontological Christology which speaks of Christ as truly God and truly man. My own tentative thoughts in this direction have sought to combine the notion of a true divine engagement with the reality of time together with the insights of a modified kenotic Christology.[12] It constitutes an attempt to find a middle way between the inadequacy of an inspirational Christology (excluding true divine participation) and the unmitigated classical account (always, it seems, in some danger of excluding a true assumption of the human).

REALISATIONS THAT THE NEW THEORY
HAS DEEP IMPLICATIONS

The difficulty of the theological endeavour is such that it has not proved possible to attain a fully articulated Christological theory. Yet the centuries of Christian reflection on the mystery of divine and human presence in Christ have brought to light profound implications which encourage us to believe that we are on the right track in continuing to wrestle with these problems. Let me illustrate this by referring to an insight that has special significance for twentieth-century thought.

The problem of evil and suffering has given rise to justified theological perplexity in every century, but it has assumed a particular significance in the century of the Holocaust. I have already discussed certain philosophical insights, such as the free-will and free-process defences (p. 14), which afford some modest rational relief by suggesting that the painful bit-

12. Polkinghorne, *Christian Belief/Faith of a Physicist*, 141–42.

terness of the world is not gratuitous but it is the ineluctable shadow side of certain other goods. That is, perhaps, of some mild help as far as it goes, but it does not go anything like far enough. The problem of suffering is not simply a rational conundrum; it is a deep existential challenge to human trust in the value and victory of goodness. It operates at a deep psychic level and it can only be met at that same profound level. The Christian response to the problem of evil and suffering is a Christological response.

In the cross of Christ we see a lonely figure, nailed to the tree, exposed to the most tortured and lingering death that Roman judicial ingenuity could devise, deserted by his friends and taunted by his enemies, experiencing also the depths of alienation from the God who had been the centre of his life, the One he knew as his dear Father, so that he cries, "My God, my God, why have you forsaken me?" (Mk. 15:34; Mt. 27:46). Christians believe that in this bleak scene we see, not just a good man caught and destroyed by the system, but the one true God who, in the taut stretching of Christ's arms on the cross, embraces and accepts the bitterness of the world that is the divine creation. The Christian God is not a compassionate spectator, looking down in sympathy on the sufferings of the world; the Christian God is truly the "fellow sufferer who understands," for in Christ God has known human suffering and death from the inside. The Christian God is the Crucified God.

Jürgen Moltmann is the theologian who has most powerfully explored and opened up for us this paradox of the God who shares the human condition, even to the point of experiencing the alienation of God from Godself. On the cross,

"Even Auschwitz is taken up into the grief of the Father, the surrender of the Son and the power of the Spirit." [13]

This profound and difficult thought meets the problem of suffering at level which its deep challenge demands. The insight of the Crucified God lies at the very heart of my own Christian belief, indeed of the possibility of such belief in the face of the way the world is. But this can only really be so if God is indeed truly present in that twisted figure on the tree of Calvary. Only an ontological Christology is adequate to the defence of God in the face of human suffering. God must really be there in that darkness. [14]

I ended my analysis of theory development in science by asserting what I have sought to defend elsewhere, namely that our theories give us verisimilitudinous knowledge of the structure of the physical world. [15] I would want to make the same critical realist claim for theology. My grounds for doing so are two-fold.

One is the stance, so natural to the scientist, that concepts that have broad explanatory power, making swathes of experience intelligible, should be expected to have ontological reference. They make sense of the world precisely because they bear some relation to the actuality of the world. They may refer to unseen and unseeable entities (confined quarks and gluons; the invisible God) but the warrant for belief in the existence of these entities is precisely that that existence provides the basis for understanding what is happening. I know

13. J. Moltmann, *The Crucified God* (SCM Press, 1974), 278.

14. Cf. V. White, *Atonement and Incarnation* (Cambridge University Press, 1991).

15. Polkinghorne, *Rochester Roundabout* (Longman/W. H. Freeman, 1989), chap. 21; see also chap. 5.

that there is much philosophical argument that rages around this point. For the present purpose I can do no more than nail my colours to the mast. Scientists, and theologians of a realist cast of mind, have one important commitment in common: they both believe that there is a truth to be found or, more realistically, to be approximated to. This belief does not entail a naive objectivity, either in subatomic physics or even less in theology. What we know of entities must conform to their nature and there is a necessarily veiled character to our encounter both with the quantum world and with God. Yet that encounter is a real meeting with something other than human thought, an exploration of what is and not just of what we choose to say.

The second reason is a particularisation of the first. The concepts we are considering cannot do the work that is needed to be done unless they have that ontological reference. A God who is just an internalized symbol of our commitment to the highest values may provide a focus for living but such a God is not the ground of hope in the face of death and beyond death. Unless there really is a God who really was "in Christ reconciling the world to himself" (2 Cor. 5:19), then the cross is no answer to the bitter problem of the suffering of the world. Indeed there could then, in fact, be no such answer beyond a heroic individual human defiance of it. Socrates would have died a nobler death than Jesus, even if he were thought to have done so in a mistaken belief in immortality.

The subject of this chapter is a particular variation on a familiar theme, one that is frequently present in the writings of those who consider the relationship between science and theology. To many of us, and perhaps especially to those whose formation lies in the sciences, it seems that there is a considerable degree of cousinly relationship between the two disciplines as each pursues its search for truth by means of the quest for motivated belief arising from their two very different domains of experience.[16]

To people who have not thought much about these matters, such a conclusion can come as a surprise. In the popular mind there is a caricature account of science inexorably progressing by means of its gain of certain knowledge, whilst theology inhabits an ivory tower whose occupants are much given to fanciful and ungrounded speculation. Neither picture is accurate, as I have tried to show. The progress of science is indeed impressive but by no means total or incorrigible. Scientists proceed as much by intuitive leaps of the imagination as by the painstaking sifting of data. The ability of Schrödinger and Heisenberg to discover their versions of quantum mechanics exhibits such creative gifts to the highest degree. Science can also live, if not comfortably at least pragmatically, with serious conceptual questions unresolved (the measurement problem in quantum theory).

Theology, on the other hand, must always seek to relate its concepts to the foundational events of its tradition and the continuing experience of its worshipping and believing community, for otherwise it would merely be creating idols. I seek to show how this grounding in experience is in-

16. M. Banner, *The Justification of Science and the Rationality of Religious Belief* (Oxford University Press, 1990); I. G. Barbour, *Myths, Models and Paradigms* (SCM Press, 1974); J. R. Carnes, *Axiomatics and Dogmatics* (Christian Journals,1982); A. R. Peacocke, *Intimations of Reality* (University of Notre Dame Press, 1984); J. C. Polkinghorne, *Reason and Reality* (SPCK/Trinity Press International, 1991); W. van Huyssteen, *Theology and the Justification of Faith* (Eerdmans, 1989).

deed the case by considering the example of Christology. At a casual glance, wrestling with the problem of how divinity and humanity could find a lodging in the same person might seem the height of metaphysical speculation. In actual fact, this persistent activity of the Church has arisen directly from the need to attempt to give a coherent and adequate account of the fact of its encounter with Christ.

Of course, the analogy between scientific and theological enquiry is not complete. Theology does not enjoy the luxury that experiment grants to science, of being able to deal with essentially controllable and repeatable experience. It has to look to the given and unrepeatable revelatory events in which God has chosen to make the divine nature known. The closest scientific analogues are cosmology's reconstruction of the unique history immediately following the big bang and biology's reading from the fossil record the story of the unique evolutionary development of life. Theological enquiry is also not simply concerned with quenching the intellectual thirst for understanding. Its insights demand response and carry implications for human conduct.

Yet, in both science and theology, the central question is, and remains, the question of truth. We shall never attain a total grasp of it but in both disciplines we may hope for a developing understanding of it.

CHAPTER THREE

Does God Act in the Physical World?

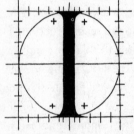N the past ten years, there has been a considerable amount of thought and speculation among those concerned with the interface between science and theology, concerning the extent to which it is possible to speak with integrity about the notion of God's acts in the world, whilst at the same time accepting with necessary seriousness what science can say about that world's regular process.[1] Many factors have made this a suitable subject for discussion.

The first is simply that it is a perennial issue on the Christian agenda. The use of personal language about God,

---

1. I. G. Barbour, *Religion in an Age of Science* (Harper and Row, 1990), chap. 8; A. R. Peacocke, *Theology for a Scientific Age*, enlarged ed. (SCM Press, 1993), chap. 9; J. C. Polkinghorne, *Science and Providence* (SPCK, 1989); *Reason and Reality* (SPCK/Trinity Press International, 1991), chap. 3; *Science and Christian Belief* (SPCK, 1994), published simultaneously as *The Faith of a Physicist* (Princeton University Press, 1994), chap. 4; K. Ward, *Divine Action* (Collins, 1990); V. White, *The Fall of a Sparrow* (Paternoster Press, 1985).

however stretched and analogical such language is rightly recognised as being, carries with it the implication of particular divine response to particular creaturely circumstance. God is not like the law of gravity, totally indifferent to context and uniformly unchanging in consequence. The Christian God is not just a deistic upholder of the world. If petitionary prayer, and the insights of a providence at work in human lives and in universal history, are to carry the weight of meaning that they do in Christian tradition and experience, then they must not simply be pious ways of speaking about a process from which particular divine activity is in fact absent and in which the divine presence is unexpressed, save for a general letting-be.

Since talk of God is inescapably analogical, talk about God's action has frequently had recourse, in one way or another, to the only form of agency of which we have direct experience, namely our own power to act in the world. I shall make two assumptions about human activity. One is that it is exercised with a certain degree of freedom; that is, our impression of choosing what to do is not an illusion. I am aware, of course, that this is a matter of philosophical contention, but I cannot here attempt to enter into that argument. For my present purpose, I shall treat human choice as being an irreducible fact of human experience. The second assumption I shall make is that we are psychosomatic unities, indivisible animated bodies, and not a dual and separable combination of flesh and spirit. Such a view sits well with our experiences of the interdependence of mind and matter (the effect of drugs or brain damage, the execution of willed intentions, our understanding that we have evolved continuously from the original quark soup of the early universe). Needless to say, I cannot solve the problem of how brain and mind relate to each other, but I look for a solution along the lines of a dual-aspect monism, a complementary account of matter in "information"-bearing-pattern, which I have tentatively and, of course, inadequately discussed on other occasions.[2] Such a stance takes our material constitution seriously but it does not capitulate to a reductionist materialism, for it asserts with equal vigour the existence of an irreducible mental pole in human nature. Bearing in mind that all conscious knowledge, even of the physical world, is appropriated mentally, such an even-handed treatment of mind and matter seems absolutely essential if we are to frame a credible account of our experience. That unconscious atoms have combined to give rise to conscious beings is the most striking example known to us of the hierarchical fruitfulness of our universe, in which there is a nesting and ascending order of being, corresponding to the transitions from physics to biology to psychology to anthropology and sociology.[3]

A further factor of considerable importance is the recognition by twentieth-century science that there are many *intrinsic* unpredictabilities inherent in the process of the physical world. If we define a mechanical system as one whose behaviour is predictable, and so in principle tame and controllable, then our century has seen the death of a merely mechanical universe. Several discoveries have brought this about.

---

2. J. C. Polkinghorne, *Science and Creation* (SPCK, 1988), chap. 5; *Reason and Reality*, chap. 3; *Christian Belief/Faith of a Physicist*, chap. 1. "Information" is being used in some highly generalised sense related to dynamic structure, which is beyond my power to specify with precision.

3. Many writers have commented on this hierarchy. A detailed and itemised discussion is given in Peacocke, *Theology*, chap. 12.

One, of course, is the well-known feature of quantum theory that permits us only to assign probabilities for the observed outcomes of quantum events. Another discovery, relating to effects operating in the macroscopic realm of classical physics and everyday occurrences, is the identification of the widespread sensitivity to minute details of circumstance displayed by those many systems whose behaviour is called "chaotic."[4] Since the slightest disturbance totally changes the dynamic behaviour of chaotic systems, caused by the exponential growth of the effects of such perturbations, the theory of chaos describes a realm of intrinsic unpredictability and non-mechanical behaviour.

This latter realisation—that Newtonian physics is not as robust as two and a half centuries of its exploitation had suggested—came as a great surprise. Our minds were unprepared because we had all been bewitched by another great discovery of Sir Isaac: the calculus. This wonderful mathematical method is precisely adapted to the description of continuous and smoothly varying quantities. Its geometrical counterparts are the well-behaved curves we can sketch with our pens upon a sheet of paper. While there are indeed such bland mathematical entities present in the patterns of the world, there are also many entities of a much more jagged character. These are the celebrated fractals, exhibiting roughly the same character on every scale of investigation, saw edges whose teeth are saw-edged, and so on down in an unending proliferation of structure that never settles to a tame unbroken line. Our mathe-

4. See J. Gleick, *Chaos* (Heinemann, 1988); I. M. Stewart, *Does God Play Dice?* (Blackwell, 1989).

matical imaginations have been greatly enlarged and enriched by this considerably expanded portfolio of possible behaviour. The world is stranger than Newton had enabled us to think.

If a clockwork universe is no longer on the scientific agenda, one must ask what is to take its place? Unpredictability, after all, is an epistemological property, simply telling us that we cannot know in detail the future behaviour of quantum or chaotic systems. Moreover, such behaviour is not totally random. An unstable atom will be able to decay only in certain specific ways and each of these options will have a quantum probability assigned to it, so that in a large collection of atoms of the same kind (a lump of matter), these different future behaviours will occur as calculable fractions of what is happening. A chaotic system is not totally "chaotic" in the popular sense, corresponding to absolutely random behaviour. Its future options converge to a certain portfolio of possibility called a "strange attractor" and it is only this limited range of contingencies that will be explored by the system in an apparently haphazard fashion. In consequence, although the detailed future behaviour of a chaotic system is unknowable, there are certain things that can reliably be said about the generic character of what will happen.

There is no logically inevitable way to proceed from epistemology to ontology, from what we can know about entities to what they are actually like. However, unless we believe ourselves to be lost in a Kantian fog—that is, unless we are condemned to groping encounter with phenomena (appearance) and we totally lack any grasp of noumena (reality)—we must suppose there to be some connection between the two. What that connection should be is a central question for philosophy and, perhaps, the central question for the philosophy of science. It can be resolved only by an act of metaphysical decision. Such an act cannot be logically determined a priori, but it can be rationally defended a posteriori, by an appeal to the fruitful success of the strategy adopted. The decision made by the vast majority of working scientists, consciously or unconsciously, is to opt for critical realism, which one could define as being the attempt to maximise the correlation between epistemological input and ontological belief. In my view, to put the point with extreme brevity, the cumulative success of science provides the necessary support for the pursuit of this strategy.[5]

In the case of the unpredictabilities of quantum theory, this has been the attitude adopted, not only by most physicists but by a great many philosophers as well. Heisenberg's uncertainty principle, which made the epistemological assertion of the simultaneous unknowability of both position and momentum, has been widely interpreted as a principle of indeterminacy, with the ontological implication that quantum entities do not possess at all times definite positions and momenta. The work of David Bohm and his colleagues in framing an alternative quantum ontology, shows clearly enough that this is not a forced move.[6] The extreme popularity of the indeterminacy interpretation has been due, I believe, not just to its chronological priority but also to a certain naturalness about an approach that allows overt epistemology to be the guide of ontological conjecture.

In the case of chaotic systems the same tendency is not apparent. No doubt a historical effect is at work here—after

5. See chapter 5.

6. D. Bohm and B. J. Hiley, *The Undivided Universe* (Routledge, 1993); see also J. T. Cushing, *Quantum Mechanics* (University of Chicago Press, 1994).

all the subject derived from the study of Newtonian equations, so that a ready-made interpretation was immediately to hand, indeed it is often called "deterministic chaos." I shall later argue that what is metaphysical sauce for the quantum goose should be metaphysical sauce for the chaotic gander. At the same time I shall explain how I believe the equations of Newtonian physics should be understood.

Let us return to the consideration of divine agency. We have seen the theological motivation for speaking of God's action and also something of the character of the physical setting of the world in which such acts would have to take place. It is time to consider what proposals have been under discussion.

A minimalist response is to decline to speak of particular divine actions and to confine theological talk to the single great act of holding the universe in being.[7] Not only is such a timeless deism inadequate to correspond to the religious experiences of prayer and of an intuition of providence but it is also interesting that it has not commended itself to those scientist-theologians who have written on these matters.[8] They do not suppose that modern science condemns God to so passive a role. Divine upholding of the cosmos, whose regular laws are understood as reflections of God's unchanging faithfulness, is part of the story of God's relationship with the unfolding history of creation, but it cannot, and need not, be taken to be the whole of that story.

Much more popular, both as an explicit theory and as a

---

7. G. D. Kaufman, *God the Problem* (Harvard University Press, 1972); M. Wiles, *God's Action in the World* (SCM Press, 1986).

8. Barbour, Peacocke, Polkinghorne, n. 1; see the discussion in J. C. Polkinghorne, *Scientists as Theologians* (SPCK, 1996), 31.

tacit understanding of what might be involved in providence-talk, has been the idea that God acts only through divine influence on people.[9] It is proposed that it is in the depths of the human psyche, rather than in the process of the external physical world, that divine agency is to be located. God's actions are those of inspiration and encouragement to human persons. A little reflection, however, soon shows that there are grave difficulties with this point of view. First, it implies that God has been an inactive spectator of the universe for most of its history to date, since conscious minds seem not to have been available for interaction with divinity until, at most, the past few million years or so of that fifteen-billion-year history. Second, and most important, if we take the psychosomatic view of human nature advocated above, then God cannot interact with the psyche without also interacting with the physical process of the world, since we are embodied beings. There is no totally separate realm of spiritual encounter, divorced from the physical/mental reality of a dual-aspect monistic world, in which providence can act. God cannot touch our minds without, simultaneously and inextricably, in some way touching our brains as well.

Process theology has sought a way round these difficulties by proposing a view of physical development in which events are the fundamental units, and all events have an experiential aspect that permits divine interaction by way of a "lure" towards a particular outcome.[10] It would, perhaps, be

9. E.g., D. Bartholomew, *God of Chance* (SCM Press, 1984), 143: divine action "in the realm of the mind."

10. See J. B. Cobb and D. R. Griffin, *Process Theology* (Westminster Press, 1976). The great exponent of process thought in relation to science and theology has been Ian Barbour.

too crude to characterize this as a panpsychic view of reality, but it certainly seeks to describe an unbroken continuum of processes within which divine interaction with a person or with a proton could both find a place, though obviously at opposite ends of the spectrum. There is then the possibility of providential interaction throughout all cosmic history, with intensification but no qualitative change, at the moment of the arrival of conscious minds on the terrestrial scene. I have two difficulties with this account of God's activity. One is physical-philosophical: I do not see that the physical world, as disclosed to scientific exploration, can be held to correspond to a concatenation of events in the manner suggested. Quantum physics involves both continuous development (the Schrödinger equation) and occasional sharp discontinuities (measurements) but it does not, to my mind, suggest the discrete "graininess" that process thinking seems to suppose. The second difficulty is theological. The God of process theology works solely through "persuasion." There is a divine participation in each event but, in the end, the event itself leads to its own completion. (It is difficult to write about process ideas without a continual lapse into panpsychic-like language.) I think this places God too much at the margins of the world, with a diminished role inadequate to the One who is believed to care providentially for creation and to be its ultimate hope of fulfilment.

An alternative strategy is to exploit rather directly an analogy between God and creation on the one hand and human beings and their bodies on the other.[11] God is then

---

11. G. Jantzen, *God's World, God's Body* (Dalton, Longman and Todd, 1984); for a critique, see Polkinghorne, *Science and Providence*, 18–22.

supposed to be embodied in the universe as we are embodied in parts of it, and to act on the whole as we do on the matter that makes us up (in whatever fashion that might be). It seems that many difficulties beset this proposal. First, the universe, though it certainly does not look like a machine, does not look like an organism either. It lacks the degree of coherence and interdependence that characterises the unity of our bodies. To put the matter bluntly, if the world is God's body, where is God's nervous system within it? Second, in our psychosomatic nature we are constituted by our bodies, and in consequence we are in thrall to them as they change, eventually dying with their decay.[12] The God of Christian theology cannot be similarly in thrall to the radical changes that have taken place within cosmic history and which will continue to happen in the universe's future. Whatever suggestiveness the idea of God's embodiment in the universe might appear to have as a metaphor, it seems that it cannot successfully function as a putative account of divine action.

It is possible, however, to seek to employ the analogical possibilities of relating divine agency to human agency in a more subtle and nuanced manner. When we act, we seem to do so as total beings. It is the "whole me" that wills the localized action of raising my arm. I am not inclined to think that this is some sort of psychological delusion produced simply by adding together discrete neuron firings in the brain and particular currents in those nerves that cause muscular contractions. On the contrary, it seems plausible that there is a

---

12. Christian hope of a destiny beyond death is expressed in terms of God's resurrection act of reconstituting us in our bodily identity in the environment of the new creation.

genuine holistic content to human agency. That would imply that there is a top-down action of the whole on the parts, as well as the familiar bottom-up interaction of the parts making up the whole. The notion of such top-down causality seems to offer an attractive possible analogy to the way in which God could interact with creation.[13] However, it is also important to recognise that, though the notion of top-down causality is motivated by our human experience of agency, it is not by itself an unproblematic or self-explanatory concept. One has to ask the question of how it may be supposed that there is room for the operation of this additional holistic causal principle within the network of physical causality established by the interactions of the bits and pieces making up the whole. In other words, to use a phrase originating with Austin Farrer, we must consider what might be the "causal joint" connecting the whole to its parts, the human self to its body, God to creation.

Farrer's own answer would be that, at least in relation to providential agency, this is a question we should decline to address, because it is beyond our human power to penetrate the mystery of divine action.[14] He writes in the tradition that speaks of God's primary agency as being at work in and through the secondary agencies of creaturely causality, in an ineffable manner which can be affirmed by faith but which is veiled from the prying eyes of human reason. Despite the venerability of this way of thinking, sanctioned by St. Thomas Aquinas and developed by many subsequent Christian thinkers, it seems to me to be a fideistic evasion of the problem. I cannot give up the search for a causal joint, though I cer-

13. Peacocke, *Theology,* 53–55, 157–60; Polkinghorne, *Christian Belief/Faith of a Physicist,* 77–79.

14. A. M. Farrer, *Faith and Speculation* (A. & C. Black, 1967).

tainly acknowledge that our actual attainments in that quest will necessarily be tentative and provisional. With the nature of human agency still mysterious, we can hardly dare to aspire to more than hopeful speculation when it comes to talk of divine agency. Yet the demand for an integrated account of both theological and scientific insight impels us to the task.

I have said that I do not expect top-down agency to be just a conglomerative effect of a lot of little bits of bottom-up interactions (in the way that the temperature of a gas is the average of the individual kinetic energies of its molecules). If holistic causality is present it must be there as a genuine novelty, and the structure of the relationships between the bits and pieces must be open enough to afford it room for manoeuvre. In some sense there must be gaps in the bottom-up account which this top-down action fills in, but those gaps must be intrinsic and ontological in character and not just contingent ignorances of the details of bottom-up process. They must be "really there" if they are to provide the causal joint for which we are looking.

Immediately there comes again to mind those widespread unpredictabilities that twentieth-century science has identified as being present in physical process. If they are to be of significance in relation to holistic causality, then they must be interpreted, along the lines already discussed, in a realist way, as being signs of actual ontological openness.

A popular site for such an exploration has been the uncertainties of quantum events.[15] Because of the almost universal

---

15. W. G. Pollard, *Chance and Providence* (Faber and Faber, 1958); articles by N. Murphy and T. Tracey in R. J. Russell, N. Murphy, and A. R. Peacocke, ed., *Chaos and Complexity* (Vatican Observatory, 1995), 289–358. Nancey Murphy is critical of my use of chaos theory. She demonstrates that epistemology does not entail on-

(but not logically necessary) tendency to give these unpredict-abilities an ontological interpretation, it seems as if there is here room for manoeuvre, space for the operation of a causal joint. The proposal is not, however, without some difficulties. Subatomic events scarcely look like promising locations for holistic causality. After all, one could hardly get more "bits and pieces" than elementary particles. It is not clear the extent to which the non-locality of quantum processes (p. 28) modifies that conclusion.[16] Moreover, the "gaps" of quantum uncertainty operate only in particular circumstances, namely in those intermittent events corresponding to acts of mea-surement. By measurement, we do not mean just observation by a person, but any record of a state of quantum process in the microworld that is obtained by an irreversible registration in the macroworld of everyday occurrence. Acts of agents are located in that same macroworld. In other words, if quantum theory does have a role to play in solving the problem of agency, it will only be because its effects are amplified in some way to produce an openness at the level of classical physics. The continuing perplexities about the quantum measurement problem remind us that we do not fully understand how the levels of the microworld and the macroworld interlock with each other. It does not seem that the proponents of divine action through quantum events have been able to articulate a clear account of how this could actually be conceived as the effective locus of providential interaction.

In these circumstances it seems worthwhile to explore whether there might not also be macroscopic phenomena that

tology (no one ever supposed it did) but she takes unquestioned the indeterministic interpretation of quantum theory, which depends upon a similar conjecture.

16. None of the authors cited in the previous note discuss this.

would lend themselves to interpretation as possible causal joints. Arthur Peacocke and I have both considered this possibility.

Peacocke's examples have been chosen from cases of dissipative systems far from equilibrium, where small triggers generate large-scale patterns of an impressive kind. Such order out of chaos provides illuminating illustrations of how structures can be formed and maintained when energy is fed into open systems, thus allowing them to swim against the tide of increasing entropy.[17] This is undoubtedly the way in which living systems are able to sustain their form in a world of change and decay. It is not clear, however, that these systems really model top-down agency. First, the character of their order is long-range pattern generated by chains of local correlations and the confining boundary conditions. That seems more sideways than top-down. Second, what is involved by way of consequence is the generation and preservation of structured pattern, whilst agency seems to require a much more open and dynamical exploration of future possibility.

The way a chaotic system traverses its strange attractor seems a more promising model for such open developments, and this has been the basis for my own suggestions.[18] We can consider the many different trajectories through the attractor's phase space (that is, the range of its future possible states) which all correspond to the same total energy. Their different forms are understood as arising from the effects of vanishingly small disturbances that nudge the system along

17. Peacocke, *Theology*, 53–55; see I. Prigogine and I. Stengers, *Order out of Chaos* (Heinemann, 1984).

18. Polkinghorne, *Providence*, 26–35; *Reason and Reality*, chap. 3; *Christian Belief/Faith of a Physicist*, chap. 1.

one path or another, the diverging characters of these different paths corresponding to the chaotic system's extreme sensitivity to perturbations.

It is this sensitivity that produces the intrinsic unpredictabilities. In a critical realist re-interpretation of what is going on, these epistemological uncertainties become an ontological openness, permitting us to suppose that a new causal principle may play a role in bringing about future developments. The character of this principle would be two-fold. First, since the paths through the strange attractor all correspond to the same energy, we are not concerned with a new kind of energetic causality. The energy content is unaffected whatever happens. What is different for the different paths through phase space is the unfolding pattern of dynamical development that they represent. The discriminating factor is the structure of their future history, which we can understand as corresponding to different inputs of *information* that specify its character (this way, not that way). Second, although the diagnostic indicator of chaotic systems is their sensitivity to small triggers, rather than this implying that we should consider them at the level where these individual small fluctuations occur, it forces on us, in fact, a holistic treatment, since the systems' vulnerability to disturbance means that they can never be isolated from the impact of their total environment.

Thus a realist reinterpretation of the epistemological unpredictabilities of chaotic systems leads to the hypothesis of an ontological openness within which new causal principles may be held to be operating which determine the pattern of future behaviour and which are of an holistic character. Here we see a *glimmer* of how it might be that we execute our willed intentions and how God exercises providential interaction with creation. As embodied beings, humans may be expected to act both energetically and informationally. As pure spirit, God might be expected to act solely through information input. One could summarise the novel aspect of this proposal by saying that it advocates the idea of a top-down causality at work through "active information." This is a phrase that Peacocke uses also. I locate the relevant causal joint in chaotic dynamics; he appears to regard God as constituting the "boundary condition" of the universe.[19]

I shall make a series of comments on this proposal, first of a scientific character and then in relation to theology.

The first scientific comment is whether one could not combine the widely acknowledged exquisite sensitivity of chaotic systems together with the widely believed openness of quantum systems, to yield a theory whose openness would result from the vulnerability of the macroscopic system to the indeterminate details of its microscopic quantum constituents. Putting it another way, macroscopic openness could be chaotically amplified quantum openness. In the end, of course, there must be a unified account combining the microscopic and the macroscopic, since there are not two physical worlds but one world encountered at these two different levels. However, the difficulties we have in understanding fully how the two levels relate to each other makes me wary of claiming an immediate synthesis. Not only is there the unsolved measurement problem to which we have already referred but also there is still considerable perplexity about what correspondence can be established between chaotic dynamics and quantum me-

19. Peacocke, *Theology*, 59-61, 161-65, 203-6. See the discussion of Polkinghorne, *Scientists as Theologians*, chap. 3.

chanics. Without attempting a detailed technical discussion, I must content myself simply to note that the nature of the compatibility of the two has not been established.[20]

The second scientific comment is that, if the proposal is correct, then at the macroscopic classical level the Newtonian deterministic equations for bits and pieces are only approximately valid as limiting cases of more subtle and flexible laws of nature in which the behaviour of parts is dependent on the setting of the whole in which they participate. This contextualism is the way in which top-down influence is brought to bear. The limit involved in obtaining the Newtonian description is obviously separability, achieved in those situations (which certainly exist but which are a subset of all possible occurrences) in which a part can effectively be isolated from the context of its whole. These are precisely the situations in which most experimental investigation takes place, since the relevant system must be capable of being treated locally and separated from its cosmic context if we are to be able to understand its behaviour. Experimental science is possible precisely because there are these cases that can be treated piecemeal, without a universal knowledge of all that is. There are many examples, however, which show that this is not universally the case for chaotic systems.[21] In our experiments we are only able to investigate thoroughly a part of what is going on.

It is important to understand what is involved in this

20. See article by J. Ford in P. C. W. Davies, ed., *The New Physics* (Cambridge University Press, 1989), 348–72. See also, however, the logic-based discussion of how classical determinism may be considered to emerge from quantum mechanics, given in R. Omnès, *The Interpretation of Quantum Mechanics* (Princeton University Press, 1994), esp. 227–34. Omnès regards this emergence as problematic for chaotic systems.

21. See Polkinghorne, *Providence*, 28–29.

proposed reinterpretation of what is often called deterministic chaos.[22] The original theory had a deterministic ontology (expressed by its Newtonian equations) but this resulted in an unpredictable epistemology. Instead of adopting the conventional strategy of saying that this shows that simple determinism underlies even apparently complex random behaviour, I prefer the realist strategy of seeking the closest alignment of ontology and epistemology (theory and behaviour) by modifying the theoretical basis along the lines proposed. This strategy then has the additional advantage of accommodating the notion of top-down causality in a natural way.

I do not doubt that reluctance to embrace the notion of flexible and contextual laws of nature stems from the fact that a theory of this kind has not yet been formulated in any detail, whilst the alternative interpretation of "deterministic chaos" (localized inflexibility with mere epistemological ignorance of determining detail) has the time-honoured equations of classical dynamics as its rigorous articulation. Recently, however, Ilya Prigogine has produced some ideas that seem to be very helpful in indicating the form that a more holistic and open dynamical theory might take.[23] He studies certain equations, such as the Liouville equation of statistical mechanics, which describe the development in time of dynamical systems. One can first look for integrable solutions of these equations, that is to say solutions which have a smooth, well-behaved character such as we considered earlier when discussing the calcu-

---

22. I am grateful to Professor R. J. Russell for a helpful conversation on the issues involved.

23. I. Prigogine, "Time, Chaos and the Laws of Physics," a lecture given in London, May 1995. I am grateful to Professor Prigogine for making the text available.

lus. These solutions turn out to have the property that they can always be decomposed into sums of contributions from definite trajectories corresponding to specific picturable behaviours of parts of the system being investigated. In other words, smooth mathematical behaviour yields a localized, bits and pieces, physics account of what is happening. It is, however, mathematically possible to enlarge the class of solutions that will be admitted, in order to include what are called nonintegrable solutions. These are not so mathematically "nice" and well-behaved—their introduction corresponds to something like a transition from smooth curves to jagged fractals. It turns out that this enlargement of the range of mathematical imagination produces possible behaviours that cannot be reduced to a sum of localized specific trajectories. A holistic account is then necessary and at the same time a rigid determinism is no longer present. Prigogine says of these additional solutions that "instead of expressing certitudes, they are associated to 'possibilities.'"

Here we are presented with a model of how it can be that Newtonian ideas, which work so well for isolable systems, are not the whole story of what is going on. The new wine of chaos theory bursts the mathematical wine skins of continuous function theory. The world is indeed stranger and more exciting than Newton imagined, even at the level of his own splendid achievements.

A final scientific comment relates to the character of causality through active information. The word "information" is being used in this slogan phrase to represent the influence that brings about the formation of a structured pattern of future dynamical behaviour. This is not the same as the registration or transmission of bits of information in the sense used by telephone engineers or, more formally, by the mathematical theory of communication. A much closer analogue is provided by the "guiding wave" of Bohm's version of quantum theory. The latter encodes information about the whole environment (it is holistic), and it influences the motion of a quantum entity by directional preferences but not by the transfer of energy (it is active in a non-energetic way). For information in the sense of the telephone engineer, there is a necessary cost in energy input, since the signal has to rise above the level of the noise of the background. For the Bohmian guiding wave there is no such energy tariff; the wave remains effective however greatly it is attenuated. I believe, therefore, that it is possible to maintain a clear distinction between energetic causality and "informational" causality, in the sense of the model under discussion.[24]

A bridge between the scientific and theological discussion can be provided by a topic in which they both have an interest: time. I have been trying to build up a tentative idea of a world whose process can accommodate our human experience that there are several kinds of causalities at work in bringing about the future. The familiar account of physics can be complemented by the presence of agencies of an holistic, and indeed ultimately purposeful, kind. The resulting picture is of a world whose interlaced causalities correspond to a true becoming, in which the sense of the passage of time is no mere psychological idiosyncrasy of human psychology but an intrinsic feature of reality, because the future is not "up there" waiting for us to arrive but we play our parts in making it as we go along our temporal way.

24. Cf. the discussion of Bohm and Hiley, *Universe*, 35–38.

It is sometimes claimed that science endorses the alternative view that the universe "is" rather than "becomes"; that its true nature is not an unfolding drama in time but the frozen history of a "block universe"; that the proper way to think of it is as a single space-time entity.[25] One could call this the atemporal view of reality. Two principal reasons are usually advanced for this conclusion.

One is that the current equations of physics provide no lodging for the concept of the present moment. The time t appears as a physical parameter but there is nothing that identifies t=0 with "now." While this is certainly the case, my response would be "so much the worse for physics!" Its inability to reproduce so fundamental an aspect of human experience is to be interpreted as indicating the incompleteness of the scientific account rather than the illusory character of our experience. Only a physical reductionist could treat this reason as decisive.

The second, more subtle, reason relates to special relativity. Different observers assess the simultaneity of separated events differently and so each constructs a different surface of simultaneity, thereby assigning a different extended meaning to "now" for distant events. Yet it is important to recognise that these judgements of simultaneity are always retrospective, assigning different characters of "nowness" to past events. Observers produce different temporal organisations of the past but this, of itself, does nothing to abolish the distinction between past and future or the moving present that

25. See the discussion in P. C. W. Davies, *About Time* (Viking, 1995), chap. 2, and Davies's own opinion in chap. 13; also the article by C. J. Isham and J. C. Polkinghorne in R. J. Russell, N. Murphy, and C. J. Isham, ed., *Quantum Cosmology and the Laws of Physics* (Vatican Observatory, 1993), 135–44.

divides them. This discrimination is preserved in the space-time formulation of special relativity in which the past and future lightcones are absolutely distinct. All observers agree about causally significant orderings.

The block universe, laid out in space and time, would correspond to how classical theology from Augustine and Boethius onwards thought of God's knowledge of creation, all cosmic history being present to the divine view "at once." The future is not foreknown by God, it is simply known by the One who is totally outside time. Since God knows things in their true reality, that would seem to imply that, for classical theology, the atemporal block universe is indeed the correct picture.

Equally, if, on the other hand, it is the universe of becoming that is the correct picture, then surely God must know it in its temporality, as it actually is. God must not just know that events are successive; they must be known in their succession. This implies that temporal experience must find a lodging within the divine nature. Of course, God is not simply temporal, in thrall to time as we are. There must be both temporal and eternal poles to divinity.[26] This notion of divine dipolarity is a gift from the process theologians that has found much wider acceptance among many other twentieth-century theological thinkers.

Logically there is a distinction between the issue of atemporal or temporal accounts of reality and the question of whether the world is deterministic or open. Yet, while there is no entailment from atemporality to determinism or from temporality to openness, there is a certain degree of natural

26. See Polkinghorne, *Providence*, chap. 7.

pairwise association between these points of view. In a deterministic universe, full knowledge of the present grants retrodictive knowledge of the past and predictive knowledge of the future. (This is the argument of Laplace's celebrated calculating demon.) Therefore an atemporal, or block universe, account is a natural way of describing a deterministic world. It is little wonder that Einstein (who wrongly associated determinism with physical reality) was so insistent that the distinction between past, present, and future is an illusion. On the other hand, classical theism has always claimed that the atemporal view does not contradict human free actions since they are not foreknown but known as they happen in God's timeless state of universally contemporaneous knowledge. My argument has been that it is unnatural to think of a world open to agencies that act to produce the future, as being one laid out complete in space and time, since it seems to be a world of unfolding becoming rather than static being.

Each account of divine relationship to time will also influence how it may seem best to think of divine action in the world. The atemporal God of classical theology has the whole of history present to simultaneous view in a way that has no analogue in human experience. Our modes of agency could, therefore, be expected in this case to be of little analogical significance in the search for an understanding of divine action. Primary causality and atemporal knowledge have a certain natural association with each other.

Yet, it is difficult to reconcile the atemporal view with a great deal of Christian thought. The religion of the Incarnation seems to imply a divine participation in the reality of the temporal, from a birth under Augustus to a death under

Tiberius. The God of both the Old and the New Testaments seems to have a deep engagement with historical process, with the becomingness of the world. Allowing for all the necessary, and sometimes unsubtle, anthropomorphisms inescapable in scripture, the God of the Bible seems far from the Boethian contemplator of the complete cosmic story.

If one is to talk of divine temporality, then one must face the question of which time is the divine time. What is God's frame of reference for the judging of simultaneity? The problem is not as acute as it might seem for, whatever solution is supposed, God is not a localised observer in the chosen frame but omnipresent within it.[27] As its time sweeps out history, God will experience every event *as, where and when it happens*, and know all such events in their correct causal interrelations (for special relativity guarantees that such interrelationships are frame independent). God's self-consistency must be supposed to preclude using omnipresence (with its power of simultaneous signalling) to subvert the divinely ordained laws of physics, so that the chosen frame will be hidden from any determination by a physical observer.[28] There is, however, a natural frame that might be associated with the Creator's cosmic knowledge of creation, namely the frame at rest with respect to the background radiation, which we use as the natural timeframe in which to define the age of the universe.

The proposal of this chapter is that human beings act in the world through a combination of energetic physical causality and active information, and that God's providen-

27. Ibid., 81–83.

28. There are several proposals in physics for frames with hidden particularity, e.g., Cushing, *Quantum Mechanics*, 189–91.

tial interaction with creation is purely through the top-down input of information. Many theological consequences flow from adopting this point of view:

(1) One of the dilemmas of talk about divine agency has been to find a path between the ineffable mystery of the claim presented by the idea of primary causality and the unacceptable reduction of the Creator to an invisible cause among competing creaturely causes (making God just a physical interventionist poking an occasional divine finger into the processes of the universe). The continuous input of active information appears to offer the opportunity of such a tertium quid.[29] It is the translation into the mundane language of conjecture about causal joints, of a long tradition of Christian thinking that refers to the hidden work of the Spirit, guiding and enticing the unfolding of continuous creation.

(2) If it is the unpredictabilities of physical process that indicate the regions where forms of holistic causality can be operating, then all such agency, including divine providence, will be hidden within these cloudy domains. There will be an inextricable entanglement—it will not be possible to itemise occurrences, saying that God did this and nature did that. Faith may discern the divine hand at work but it will not be possible to isolate and demonstrate that this is so. In this sense, the causal joint is implicit rather than explicit. The veiled presence of God, discreetly hidden from contact with finite human being, may be held to require divine actions to be thus cloaked from view. The theological assessment of the balance between what God does and what creatures do, is the old problem of the balance between grace and freewill, now being considered on a cosmic scale.

29. I am grateful to Professor R. J. Russell for a helpful conversation on this point.

(3) There are, of course, predictable aspects of natural process that the divine consistency can be expected to maintain undisturbed as signs of God's faithfulness. The succession of the seasons and the alternations of day and night will not be set aside.

(4) Considerations of divine consistency lead us to expect that in comparable circumstances God will act in comparable ways, though the infinite variety of the human condition means that no simple lessons can be drawn from this about individual human destinies. In unprecedented circumstances, it is entirely conceivable that God will act in totally novel and unexpected ways. That is how I try to understand claims about divine miracles, a subject which lies outside the humdrum limits of the present discourse,[30] but one which is of central importance to a Christian thinker because of the pivotal role played by Christ's resurrection.

(5) If the physical universe is one of true becoming, with the future not yet formed and existing, and if God knows that world in its temporality, then that seems to me to imply that God cannot yet know the future. This is no imperfection in the divine nature, for the future is not yet there to be known. Involved in the act of creation, in the letting-be of the truly other, is not only a kenosis of divine power but also a kenosis of divine knowledge. Omniscience is self-limited by God in the creation of an open world of becoming.

The God of the atemporal view knows all history and interacts with it in a unified, if mysterious, fashion. The God of temporal process does not yet know the unformed future and interacts with history as it unfolds, responding to its development in the way so often described anthropomorphically

30. Cf. Polkinghorne, *Providence*, chap. 4.

in the Bible. The One is the Composer of the whole cosmic score; the Other is the Great Improviser of unsurpassed ingenuity (in Arthur Peacocke's striking phrase) of the cosmic performance. It is clear that the God of temporal process is the more vulnerable in relation to creation than is the atemporal God of classical theism. The converse of that is that it seems that the atemporal God presents greater difficulties for theodicy than does the God of temporality. The discords in the score are simply there in the former case, rather than arising from the uncertain clashes of contingent process. The necessary precariousness of love's creative acts, so movingly described by W. H. Vanstone, finds its most natural expression in terms of the unfolding temporal view.[31]

It appears that there are certain theological concepts, which though not logically locked together, tend to constellate in association with each other. In one group are atemporality, primary causality, divine impassibility, an inclination towards determinism, and an emphasis on divine control. In the other group are temporality, top-down causality, divine vulnerability, an inclination towards openness, and a recognition of creaturely self-making. While the former group is the one endorsed by much of the classical Christian tradition, it is the latter group which I believe accords best with late twentieth-century thought, both scientific and theological.

This chapter has dealt with great matters in a way that is necessarily limited, inadequate, conjectured. We are a long way from a full understanding of our own powers of agency, let alone how it is that God works in the world. Still, the task

31. W. H. Vanstone, *Love's Endeavour, Love's Expense* (Darton, Longman and Todd, 1977).

is a necessary one, and I think one can draw some modest encouragement from the drift of the argument. The dead hand of the Laplacean calculator, totally in control of the sterile history of his mechanical universe, has been relaxed. In its place is a more open picture, capable of sustaining motivated conjectures that can accommodate human agency and divine action within the same overall account. Modern science, properly understood, in no way condemns God, at best, to the role of a Deistic Absentee Landlord, but it allows us to conceive of the Creator's continuing providential activity and costly loving care for creation.

CHAPTER FOUR

The Continuing Dialogue
Between Science and Religion

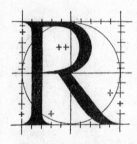ELIGIOUS thought and scientific thought have been interacting seriously with each other ever since the rise of modern physics provided the mature second partner to participate in the dialogue. Men like Galileo and Newton, whatever problems they may have had with Christian authority or Christian orthodoxy, were people for whom religion mattered. Hence the popular seventeenth-century concept of reading the two books that God had written: the book of nature and the book of scripture. The subsequent history of that mutual interaction is complex, and it is not susceptible to simple characterisation, either in terms of conflict or of harmony.[1]

The past thirty years or so have been a particularly active phase in this conversation between the two disciplines. One might, perhaps, date the beginning of this period to 1966,

1. See J. H. Brooke, *Science and Religion* (Cambridge University Press, 1992).

with the publication of Ian Barbour's *Issues in Science and Religion*, a judicious summary of matters to be considered.[2] For an academic generation this was the text that set the scene for the starting student. Barbour placed considerable emphasis on the historical story, for the events associated with Galileo and Darwin were still seen by many as representing critical (and for religion, discreditable) moments of significance. More careful and balanced scholarship enables us today to perceive the complexity of those times, in which scientists and religious thinkers alike wrestled with the difficulties and unresolved problems attendant upon periods of great intellectual change and when both kinds of participant were to be found on both sides of the argument. Barbour's work has played its part in bringing about this juster assessment. Only in the media, and in popular and polemical scientific writing, does there persist the myth of the light of pure scientific truth confronting the darkness of obscurantist religious error. Indeed, when one reads writers like Richard Dawkins or Daniel Dennett, one sees that nowadays the danger of a facile triumphalism is very much a problem for the secular academy rather than the Christian Church.[3]

Three other topics were high on Barbour's agenda. One was a methodological and philosophical revaluation of science, rejecting the poverty of positivism and drawing out science's kinship with other forms of rational enquiry. A second topic was the rejection of reductionism by maintaining the claim that the whole is more than the sum of its parts, that human beings are not "nothing but" complex collections of

2. I. G. Barbour, *Issues in Science and Religion* (SCM Press, 1966).

3. R. Dawkins, *River out of Eden* (Weidenfeld and Nicholson, 1995); D. Dennett, *Darwin's Dangerous Idea* (Simon and Schuster, 1995).

elementary particles. The third topic was an acceptance of evolutionary biology and the positive evaluation of its consequences for a doctrine of creation, representing God as acting through *creatio continua*. These themes were taken up and expanded from a biological point of view by Arthur Peacocke in *Creation and the World of Science*.[4] I have also sought to add to them such insights as come naturally to someone whose scientific experience is in fundamental physics and to emphasise the possibility of a suitably modest revised and revived natural theology, of the kind sketched in the first chapter.

Philosophy, creation, and natural theology form appropriate frontier regions in which science and theology can encounter each other. Discussion will always continue in these areas, but I think that the significant developments of the next thirty years are likely to lie elsewhere. But the character of the discussion will change, because the participants, and also the topics, will have to be enlarged in their scope.

Too much of the running to date has been made by those whose formative experience lies in physical science. (Even Peacocke is a *physical* biochemist.) We desperately need the participation of more biologists, more practitioners of the human sciences, and more theologians.

At the moment the biological world, particularly in its members who work with molecules rather than organisms, displays notable hostility to religion, at least in the writings offered to the general educated public. (It is a curious cultural fact about our society that, though it would be considered im-

---

4. A. R. Peacocke, *Creation and the World of Science* (Oxford University Press, 1979).

proper for a believing scientist to exhibit that belief explicitly when writing for the lay public about science itself — as opposed to writing books explicitly about science and religion — it is apparently perfectly all right for the atheist to press unbelief in a similar scientific context.)

I think two effects produce this hostility. One is that biologists see a much more perplexing, disorderly, and painful view of reality than is presented by the austere and beautiful order of fundamental physics. It is absolutely essential that religion takes heed of the biologists' challenges in this respect and makes a candid response to the problems represented by the cruciform nature of our world.[5] There is, however, a second effect at work of much less intellectual respectability. Biology, through the unravelling of the molecular basis of genetics, has scored an impressive victory, comparable to physics' earlier elucidation of the motions of the solar system through the operation of universal gravity. The post-Newtonian generation was intoxicated with the apparent success of universal mechanism and wrote books boldly proclaiming that man is a machine. It seems to me that a similar intellectual inebriation influences some biologists, only too ready to assert that we are nothing more than genetic survival machines. In any discipline, you solve the clockwork problems first (they are the easiest) but, as physics has found out, there is always more to the story than the ticking of clocks. I am sure that biology will eventually make the same discovery, but for the moment many of its adherents have succumbed to

5. See H. Rolston, *Science and Religion* (Temple University Press, 1987), chaps. 3, 7.

an ill-judged reductionist triumphalism. However, the madness is not universal, and we must hope for more biological colleagues to be prepared to participate in the interaction between science and theology.

Even more important will be the cooperation of those trained in the human sciences. I was encouraged when my colleague Fraser Watts, who has been a clinical and experimental psychologist, was appointed Starbridge Lecturer in Theology and the Natural Sciences at Cambridge University. Next to the deep mystery of the divine nature, the mystery of the human person is of central significance for the whole discussion, since scientific and religious concerns intersect most clearly in our embodied nature, embedded in the physical world but transcending a merely reductive physicality. We take our stand in the order of being somewhere between ape and angel.

The most grievous absence from the conversation is that of the theologians. Their presence, in a sustained rather than a merely occasional way, is earnestly desired as part of future developments. I see the difficulties for them. Because God is the ground of all that is, in some second order sense all human knowledge is the concern of the theologian. But books are many and life is short, and much science is formidably technical in appearance, although many of its concepts are capable of at least partial expression in lay language. Moreover, much twentieth-century theology has been either fideistic (Barth) or existential (Bultmann) in tone, conducted from within ghettoes walled off from scientific culture.

There have been some honourable exceptions to this policy of keeping theology at a distance from science. Thomas Torrance has sustained a life-long interest in scientific mat-

ters.[6] One emphasis in his thought that is particularly congenial to the scientific mind is the insistence that components
of reality, from electromagnetic fields to God, are known in
ways that accord with their natures and that we cannot determine beforehand what these epistemological modes will be.[7]
Torrance likes the metaphor of listening, of being receptive
to the impact of reality as it comes to us as a gift. Yet his scientific heroes are James Clerk Maxwell and Albert Einstein,
the last of the ancients rather than the first of the moderns,
and we do not find in his thought, for example, much engagement with the veiled elusiveness of the quantum world.

Jürgen Moltmann has evinced a desire to take modern
science into account, but so far the execution has fallen short
of the promise. In relation to a discussion of the "Space of
Creation," he acknowledges only a future willingness: "I hope
to be able to develop this subject further elsewhere in the
light of new scientific conceptions about the space-time continuum."[8] General relativity is not something one can master
in a week, so a certain caution is to be applauded, but it is possible to talk to those who are already professionally acquainted
with the subject. It is certainly not possible to talk seriously
today about space without taking relativity into account.

Wolfhart Pannenberg's major writing in this area has
been concerned with the philosophy of science, rather than
with science itself, and with a kind of philosophy, moreover,
which seems much more concerned with the social sciences

6. T. F. Torrance, *Theological Science* (Oxford University Press, 1969); *Divine
and Contingent Order* (Oxford University Press, 1981).

7. See the discussion in J. C. Polkinghorne, *Science and Christian Belief*
(SPCK, 1994), simultaneously published as *The Faith of a Physicist* (Princeton University Press, 1994), 32–33.

8. J. Moltmann, *God in Creation* (SCM Press, 1985), xiv; see also chap. 6.

than with anything that would be immediately persuasive to a natural scientist.[9] When Pannenberg considers the content of science, two particular themes emerge.[10] One is a preoccupation with a rather old-fashioned concept of inertia, which Pannenberg seems to treat as if it were a kind of burdensome constraint on the freedom of creation, or even of the Creator. While modern physics gives an important role to conservation laws, there seems no need to make them carry such a load of metaphysical freight. They are consequences of the symmetries of creation and can easily be understood as expressions of the Creator's will rather than impositions on it. Fruitful physical process requires a degree of stability as well as a degree of flexibility.

The second theme is the idea of a field, which Pannenberg supposes, in its spatially spread out character, to offer a metaphor, or even perhaps more than a metaphor, for omnipresent Spirit. To a physicist this seems rather quaint. A field is just a dynamical system with an infinite number of degrees of freedom (ways of changing) rather than a finite number. This implies that it is described by partial differential equations rather than ordinary differential equations, but a classical field is a perfectly mechanical and deterministic system, the carrier of energy and momentum. It is, moreover, a *local* entity, that is to say it is not an integrated unity but different bits can be changed in different ways in whatever manner one chooses to implement. Its utility as a model for Spirit seems distinctly limited. There is semantic danger in transferring terms across disciplines without careful consideration of appropriateness.

9. W. Pannenberg, *Theology and the Philosophy of Science* (Darton, Longman and Todd, 1976).

10. W. Pannenberg, *Towards a Theology of Nature* (Westminster, 1993).

It is scarcely surprising that theologians often do not achieve great sophistication or insight when they turn to science. We scientists who take a serious interest in theology are only too open to the charge of a converse lack of skill and knowledge when we venture into matters theological. I know what it is like to be a professional theoretical elementary particle physicist, with the many years of apprenticeship and engagement with the subject and its community which the attainment of that status demanded, and I know that I shall never become a fully professional theologian in the same sense. Life is too short for that. Reviewers of one sort or another can always chide the scientist-would-be-theologian for not being completely at home in the *Summa Theologiae* or not having read the whole of the *Church Dogmatics*.

The moral is certainly not that we should all return to the comfort and safety of our professional home grounds. Interdisciplinary work is both essential (for, in the end, knowledge is one) and risky (for we must all venture to speak on topics of which we are not wholly the master). We must attempt a bit of intellectual daring and, above all, we have to be prepared to listen and learn from each other, showing mutual tolerance and acceptance in doing so. I do not yet see a dialogue of this kind taking place between mainstream theologians and mainstream scientists, but I fervently hope it will be one of the leading developments of the next few years.

The topics of that broader conversation will not be confined to conventional borderline questions such as those associated with creation and natural theology. A deeper mutual interpenetration will have to take place, and indeed this is already happening to some extent. The engagement with questions of divine action, discussed in the third chapter, is one sign that this is the case. One can also see the potential fruitfulness of a discussion between informed theologians and informed scientists, in which the nature of time and the nature of God's relation to temporality would be carefully considered, with each side paying respectful attention to the insights and concerns of the other.

Three of us scientist-theologians have in recent years given Gifford Lectures.[11] Our offerings differ in style and content but all three, with some charitable generosity accorded to them, might be described as "mini-systematic theologies"; that is, they are attempts to articulate Christian belief in ways that seem natural and congenial to the scientific mind. I characterise this approach as being "bottom-up thinking." Scientists know that reality is strange and surprising; they are not tempted to make current understanding the measure of future insight. Rather than supposing that we know beforehand what reasonable belief is going to be like, scientists are content to be open to new possibilities, provided only that novel ideas are grounded in, and motivated by, new experiences of a kind that requires such an expansion of our intellectual horizons. We look to evidence for what we are asked to believe. Bottom-up thinkers proceed from the basement of phenomena to the superstructure of theory. Top-down thinkers somehow seem to start at the tenth floor and to know from the start what are the general principles that should control the answers to the enquiry. Many theologians appear to the scientists to be of the top-down variety. As they discourse on the immanent Trinity, one wonders how they know what they are claiming. Bottom-

11. I. G. Barbour, *Religion in an Age of Science* (Harper and Row, 1990); A. R. Peacocke, *Theology for a Scientific Age*, enlarged ed. (SCM Press, 1993); Polkinghorne, *Christian Belief/Faith of a Physicist*.

up thinkers prefer to stick with the economic Trinity.[12] Their method is *analogia entis* rather than *analogia fidei*, the appeal to experience and reason rather than to a superior source of knowledge.

When scientists venture to offer their approaches to Christian belief, it is important to recognise that we are not simply engaged in an apologetic exercise, trying to make the faith appear acceptable in a scientific age. I certainly do not despise that task, but the external defence of Christianity will only be effective if it has also the internal characteristic of being an exploration of Christian truth, pursued with integrity and the desire for understanding. Of course, scientific formation in no way provides a uniquely effective manner in which to tackle the task of theology. At best, the activity has its place within a spectrum of approaches that present the opportunities and limitations inherent in a variety of different perspectives on the theological enterprise. The comparison here is with movements such as liberation theology and feminist theology, which also take off from their particular initial points of view. All these insights are both useful and limited, and I claim that a science-and-theology approach has its place among them in this modest fashion. It is interesting, and somewhat sad, that the second volume of David Ford's valuable survey of modern theology,[13] which is devoted to particular lines of approach to the subject (through history, philosophy, black theology, Asian theology, etc.), initially found no place for an approach through science.

12. In Christian thought, the immanent Trinity is God in the divine nature itself; the economic Trinity is God's self-manifestation to creation.

13. D. Ford, ed., *The Modern Theologians*, 2 vol. (Blackwell, 1989). This omission has been rectified in a new edition.

This scientific avenue into theological thinking will seek to give due weight to science, but it would be fatal to allow it to become a scientific take-over bid, affording no more than a religious gloss on a basically naturalistic account. The God and Father of our Lord Jesus Christ is far from being the God of Spinoza, *deus sive natura*, a cypher for the rational order of the universe. That was Einstein's God, but it is certainly not mine.

There remains the question of the degree of accommodation required of the historic faith in its expression in an age of science. Barbour, Peacocke, and I, all firmly resist the subordination of theology to science, but there are discernible differences between us concerning the extent of the revision that is called for in pursuing the subject today. I have suggested that there is a spectrum of response running from assimilation to consonance.[14] The assimilationist seeks the most immediate and accessible correlation between scientific and religious thinking. Jesus Christ will still be accorded a preeminence, but this will be understood in the functional and evolutionary terms of a "new emergent," Christ as the pioneer of what humanity can become under the guidance of divine inspiration. The consonantist, on the other hand, while wishing to ensure that theological understanding is consistent with what science tells us about the structure and history of the physical world, will insist that theology is as entitled as science to retain those categories which its experience has demanded that it shall use, however counterintuitive they may be. Jesus Christ will continue to be understood in the incarnational terms discussed in the second chapter. I am such a

14. J. C. Polkinghorne, *Scientists as Theologians* (SPCK, 1996), chap. 7.

consonantist; my colleagues seem to display assimilationist tendencies to differing degrees.

This is a debate that will continue, and, once again, the participation of our theologian colleagues is essential. Underlying it is a contrast between the way in which science and theology relate to their past. The cumulative character of scientific understanding means that scientists sit light to the work of the generations that preceded them. Present research is built on past achievement (Newton said that he had stood on the shoulders of giants), but once we have gained the new height we can kick away the ladder that got us there. I do not need to read the *Principia*. Moreover, Isaac Newton, towering genius though he was, knew far less about the universe than does the average well-instructed physics Ph.D. of today. Religious understanding, however, does not increase with time in that linear kind of way. The theologians of the past centuries still convey to us insights that are uniquely their own. The conversation with the Fathers, with Aquinas, Luther, and Calvin, will never come to an end. The consonantist is keenly aware of that diachronic discussion, just as the assimilationist is keenly aware of the synchronic encounter with contemporary culture. Both will agree that there are some classical Christian doctrines that need restatement in an age of science and whose reconsideration will surely be part of the agenda of future development. I turn briefly to two of these.

The scale of theological thinking, in both space and time, still remains domesticated and anthropocentric. When theologians speak of the "world," they usually do not mean the universe but our local planet. When they talk of history, it is mostly the few thousand years of human cultural development that they have in mind. When they talk of the future, it seems to stretch only a few centuries onward. The ghost of Archbishop Ussher has not been wholly exorcised from theology. This means that some questions relating to cosmic beginnings and cosmic endings require further discussion.

Concern with beginnings scarcely needs to focus yet again on the tired issue of big bang cosmology. Popular science writers, who like to garnish their wares with references to God, still seem to find it difficult to grasp that the doctrine of creation is concerned with why the world exists, and continues to exist, rather than how it all began. Yet the rest of us know that theology is concerned with these ontological questions and that it gains little from science's fascinating, but largely theologically irrelevant, talk of temporal origins. Much more important is that event which is surely the most significant in cosmic history to date—the dawn of consciousness. From the theological point of view this raises the acute question of how we are to understand the Christian doctrine of the Fall.

In the sense of contemporary experience it seems straightforward. One recalls Reinhold Niebuhr's remark that original sin is the only empirically verifiable Christian doctrine! You have only to look around—or within—to see the slantedness of human nature, which frustrates human hopes and perverts human desires. Yet we can no more believe that this is the entail of a single disastrous ancestral act than we can believe that there was neither death nor thistles in the world before our forebears took that fatal step. It has long been understood that the powerful tale of Genesis 3 is to be understood mythically rather than literally. In part it portrays life as we now experience it, but that recognition does not remove the question of how these things came to be in God's supposedly good creation.

Clearly consciousness is possessed by some of the higher animals but it seems likely that the further power of self-consciousness, with its concomitant ability to form expectations and plans for the future, only dawned with the evolution of the hominid lines leading eventually to Homo sapiens. As that self-awareness developed, I suppose that a corresponding spiritual awareness of the presence of God also became part of the experience of these living beings. One can conceive of a struggle in the hominid psyche between the pole of the self and the pole of the divine, resolved by a turning from God and a concentration on the creature as all-sufficient, a succumbing to the temptation, whispered in Eve's ear by the serpent in that powerful ancient story, to assert human autonomy over creaturely dependence, to believe "you will be like God, knowing good and evil" (Gen. 3:5). In Luther's phrase, humanity became *incurvatus in se*. At what stage in hominid development, and over what period of time, this inversion upon the self took place, I do not know. That it has taken place seems confirmed by the contemporary human condition. It is in these terms that one can try to construct a contemporary doctrine of the Fall.

There was death in the world long before there were our human precursors. After all, it was the extinction of the dinosaurs that gave us mammals our evolutionary chance. But the Fall, as I have described it, turned death into mortality. Self-consciousness made us aware of our transience—we could foresee our deaths—and alienation from the God who is the eternal ground of hope, turned that recognition into sadness and bitterness. In a similar way, the problems of living, symbolised by thorns and thistles, became causes of frustration and the expense of spirit.

The second question is that of endings. It is an important theme in much contemporary theology to speak of God as End, and of true significance as only to be revealed in the future by what creation becomes.[15] Yet science tells us most surely and clearly that the end of present physical process lies in futility, the bang of cosmic collapse or the whimper of cosmic decay. Carbon-based life will have its day and then it will disappear for ever. I mentioned these problems in the first chapter. A credible eschatology, which takes account of the eventual death of the universe and looks beyond it to God's new creation, is surely an indispensable component in realistic Christian thinking. I have attempted elsewhere to outline the form that such an eschatology might take.[16] This is not the place to recapitulate those thoughts, speculative but motivated as they are, but to emphasise that Christianity must not lose its nerve as to a true hope and an everlasting destiny. Theology must summon its resources to continue the rational discussion of the promise of God's enduring faithfulness. The subject transcends what science can say, but theology's treatment of it must be consonant with our limited knowledge of present and future possibility.

The continuing development of the interaction between science and theology cannot simply be contained within the limits of Christian discourse alone. The other great faith traditions of the world must also be involved and their very existence, in their stability and unreconciled diversity, poses a profound problem for an ecumenical theology, which neither denies to the traditions their roles as carriers of authentic

15. See, e.g., the writings of R. W. Jenson, J. Moltmann, W. Pannenberg, and T. Peters.

16. Polkinghorne, *Christian Belief/Faith of a Physicist*, chap. 9.

spiritual experience nor seeks to eliminate their incompatibilities by reducing each to a lowest-common-denominator account, unrecognizable to its adherents. The particularities of the world faiths contrast strongly with the universality of scientific understanding, which is as acceptable in Banares as it is in Jerusalem, in Mecca as it is in Kyoto.[17]

Centuries of dialogue lie ahead of the world faiths. The initial meeting grounds will need to be as accessible and unthreatening as possible. Asking each tradition how it responds to scientific understanding and how it relates to science's account of cosmic history would seem to present the possibility of an oblique approach to mutual encounter through a meeting with a relevant third party. In simple terms, one perceives a good deal of commonality between Judaism, Christianity, and Islam (whose adherents historically played major roles in the growth of scientific learning) on this issue of relating to science and of acknowledging the structured reality of the physical world, but greater perplexities seem present in relation to Hinduism and Buddhism (despite the contemporary flourishing of science in countries that are among the heartlands of these traditions).

Another aspect of developing thought, going beyond specifically Christian concerns, relates to how all people of goodwill should seek to tackle the moral problems posed by the growth of science. How can we find responsible and ethical uses for those new possibilities for human intervention continually made available through the technological exploitation of scientific discovery? How can we respect and preserve the environment that belongs as much to future gen-

17. Ibid., chap. 10; see also Polkinghorne, *Scientists as Theologians*, chap. 5.

erations as to the present? The ethical snare for the scientist is to get so caught up in the excitement of research that there is never time to ask where it is going and to what end. Not everything that can be done should be done. The technological imperative must be tempered by the moral imperative. All new discoveries are "falls upward," the enlarged powers thus obtained containing the potential both for good and for ill. In seeking to make wise judgements, we need to encourage opportunities for responsible debate rather than occasions for the confrontation of single-issue pressure groups. Though there are guiding principles, the problems are usually too complex for a rule-book approach, and careful discussion of particular cases is likely to be a necessary source of insight. The expertise of scientists is an indispensable resource, but decisions cannot be left to them alone, and society must continually ask them the question of whether they have fully considered the consequences of what they are doing.

I cannot claim that Christian thinking on these issues has always been careful or well argued. The World Council of Churches has more than once issued documents that are simplistic and misleading in tone about issues that required much more balanced and expert consideration. Suspicion of the scientific community has sometimes led to foolish rejection of sensible advice. The science-and-theology community itself has so far devoted comparatively little attention to ethical matters, though Ian Barbour has made a start.[18] Much work remains to be done in the future. I believe that an ethically respectful treatment of nature ultimately requires undergird-

18. I. G. Barbour, *Ethics in an Age of Technology* (HarperCollins, 1993).

ing by a theology of nature, for the ground of a reverence for life, and of a concern for future generations, lies in our creaturely status. The world is the gift of the divine Creator, not the construct of a human exploiter. We do not possess anything that we have not received and that we will not be called on to hand on to those who follow us.

This developing mutual interaction certainly holds out the promise of gains for theology. Can it also be expected to be of use to science itself? One should not expect the potentialities for further illumination to be the same for the two disciplines. Science is a first-order activity, bringing intellectual reflection to bear on our experience of the physical world. Theology also has its first-order aspect, as it reflects on religious experience, but a good deal of its activity is second-order in character, as it seeks to integrate the findings of first-order investigations into a single unified metaphysical account. This is the mode in which theology is operating when it pursues a consonant relationship between its insights and those of science. The two disciplines, therefore, are not connected with each other in a symmetrical fashion.

Nevertheless, there may well be a point of understanding on which the concerns of both theology and science converge. In the third chapter, we considered the problems of divine and human agency and suggested that there might be holistic causal principles of a pattern-forming kind, whose operation was summarised in the phrase "active information." I believe that this idea could find application outside the sphere of understanding the acts of agents. Science should consider the possibility that there are also holistic laws of nature of this pattern-forming character that might well have played their part in the astonishing drive to complexity manifest in cosmic and terrestrial history.[19]

The classical neo-Darwinian explanation of the development of life on Earth, including the rapid expansion of the hominid brain over a period of a few hundreds of thousands of years, has been to assign it totally to the sifting and preservation through the process of natural selection of the effects of small random genetic mutations. No reasonable person doubts that this is a component in the history of life but that it is the sole and totally adequate cause of all that has happened is simply an article of blind belief. It is a scientifically interesting question to ask whether there might be more to the story than has been told. In my own mind, the magnitude of the changes involved in the time scales available encourages the thought that there might be more to discover. Holistic laws of nature, of the novel kind suggested, represent one possibility that is worth considering.

The suggestion can be explored using some concepts dear to the heart of that arch-Darwinian Daniel Dennett.[20] He likes to picture the development of life as the process of exploring "Design Space." This is a vast abstract array of conceivable beings (a space of possibilities, vastly infinite dimensional in a mathematical sense), representing all the forms that life might have taken (the phenotypes resulting from the genotypes corresponding to all imaginable viable skeins of DNA). Of course, in reality, over the few-billion-year history of terrestrial life, only a tiny fraction of this vast potentiality has actually been explored and realised, but we can picture the

19. See P. C. W. Davies, *The Cosmic Blueprint* (Heinemann, 1987).
20. Dennett, *Dangerous Idea*, chaps. 5, 6.

history of life as having been the progressive colonisation, and often eventual abandonment, of locations in that particular fragment of Design Space. The way life has spread out has not been totally random, for there is a metric on Design Space, a pattern of hills and valleys which directs development in one way and not in another. It is easier to go downhill than uphill, so some options are more likely than others.[21] The only source of that metric which Dennett's Darwinian orthodoxy permits him to consider is the "fitness landscape," resulting from the differential advantage conferred by superior survival value. I suggest the possibility of a further element in the metric, resulting from the operation of holistic laws of nature that encourage the formation of certain kinds of pattern and inhibit the formation of others. This would be an extra effect, playing its part in driving the "optimistic arrow" of fruitful change.

Here is one simple example of how such laws might function, drawn from the known pattern-forming propensities of systems far from equilibrium (see p. 61). The particular phenomenon in question is called Bénard instability. If fluid is confined between two horizontal plates with the bottom plate hotter than the top, and if the temperature difference is sufficiently great, convection currents are set up transferring heat energy from bottom to top. This convective motion takes a very specifically ordered form, corresponding to an hexagonal pattern of convection cells within which the fluid motion is contained. In purely geometrical terms a pattern of square cells would be an equally "possible" kind of order

---

21. I choose the convention that the "lower" points are the more favoured. This is the reverse of the picture used by most biologists who like to speak of peaks of fitness.

(both squares and hexagons completely tile the plane), but there is in nature a preference for the hexagonal arrangement. The generation of order from chaos in this particular case creates a metric that directs the development of pattern of an hexagonal, rather than a square, kind. This very simple example may be suggestive of much more complex and interesting possibilities about how order is selectively generated in the physical world.

In fact, work of this kind is already being undertaken. Stuart Kauffman has been concerned with the consequences for evolutionary biology of the generation of structure at the borderline between chaotic and regular systems. He suggests that at the edge of chaos "the typical, or generic, properties of such poised systems emerge as potential ahistorical universals in biology."[22] These insights are intended to supplement, not supplant, natural selection. To refuse to consider them would be foolish for "it would be shortsighted to ignore the possibility that much of the order we see in the biological world reflects inherent order."[23] There may be much more potentiality for complex structure built into the very fabric of the universe than we have yet fully realised. Life has arisen through the interplay of chance and necessity, and any account which concentrates on historical contingency to the neglect of ahistorical lawful order is unbalanced and inadequate.

Chaos theory and complexity theory are still in their infancies. We are at the "natural history" stage of investigating particular cases, studying behaviour generated by means of extensive computational exercises. There are intriguing hints

22. S. Kauffman, *The Origins of Order* (Oxford University Press, 1993), xv.
23. Ibid., xvi; cf. B. Goodwin, *How the Leopard Changed Its Spots* (Weidenfeld and Nicholson, 1994), with its emphasis on the role of organisms.

of deep general structure lying beneath the behaviour of specific examples, but there is not yet an adequate theory to pull it all together.[24] The metaphysical message is ambiguous. Either it can be read in a reductionist fashion as indicating that the rich structure of the world is merely the elaboration of a fundamental simplicity, or it can be read in a holistic way as indicating the inadequacy of a mechanical view for the task of capturing the subtle and exquisitely sensitive patterning of actual behaviour. I prefer the latter metaphysical strategy as the one more promisingly compatible with human experience (see chapter 3). Such holistic and interconnectional concepts are congenial to theological thinking, as exemplified, for instance, by much Trinitarian discussion that emphasises relationship (communion) as the ground of being.[25]

The intellectual world in which these future discussions will take place is characterised in the minds of many contemporary thinkers as coming after the great intellectual movements of the preceding three hundred years, the period in which science came to birth and fruition. Now we are "post-Enlightenment," "post-liberal," "post-modern." Amid the ruins of the Cartesian search for certainty, science and theology can proclaim, in their different ways, that there still can be found a reliable understanding of reality, verisimilitudinous rather than absolutely true, attained through the exercise of creative human powers rather than by logical de-

24. Examples of significant straws in this mathematical wind are Mitchell Feigenbaum's discovery of universality in the way that cycle-doubling generates chaos in a wide class of systems, and Stuart Kauffman's "law" that the number of stable cycles in a Boolean network with $K=2$ connectivity is proportional to the square root of the number of elements.

25. See Polkinghorne, *Christian Belief/Faith of a Physicist*, 154–56.

duction, backed by experience but not simply read out of it. In between, on the one hand, the certainty to which the Enlightenment aspired, and on the other hand, the relativism that present-day deconstructive critics proclaim, there lies the critical realist approach to knowledge, ever open to correction but persuasive that its power to make sense of experience derives from its correlation with reality. Both science and theology offer support to this middle way of intellectual enquiry (see chapter 5).

Most crystal balls are pretty cloudy, and it is not easy to see far into the future. Yet I think we can expect, and hope for, developments along the lines I describe here. The new programme of activity will be characterised by:

(1) wider participation by theologians and by experts in an extended range of scientific disciplines, with the human sciences playing a particularly significant role;

(2) wider participation from the world faith traditions, enriching the science and theology dialogue and providing also an unthreatening ground for ecumenical encounter;

(3) recognition that the bottom-up habits of thought, natural to the scientist, have their role to play in the discourse of theology and that this can lead to helpful revision of traditional concepts where that is necessitated by the advance of knowledge;

(4) recognition that holistic and relational concepts are coming to play an increasing role in science and that this can be correlated with theological insights of being in communion;

(5) facilitating a serious discussion of ethical issues which will maintain a balance between the analysis of the experts and the moral understandings of society;

(6) maintaining the critical realist approach to knowledge

that recognises that understanding requires recourse to acts of bold and motivated commitment which yield, despite their logical precariousness, an increasingly verisimilitudinous grasp of reality.

The consequence of this programme will not be for the academic world alone, but we must be prepared to make efforts to propagate these deeper understandings in the wider world as well as within the academy. Both science and theology are marginalised by much of the predominant culture of the Western world, and they could well share a common interest in how to lay their insights more successfully before the general public.

That theology is thus side-lined would be a common perception; it may seem a more surprising judgement about science, in what is frequently asserted to be a scientific age. Consideration of media coverage, the level of general knowledge, and what it is considered interesting and acceptable to raise in polite conversation at the dinner table, soon convince one, however, that contemporary engagement with science is, in the main, extremely shallow. When I give public talks about science-and-theology, I am as much a missionary for science as for religion. I recently was taking part in a panel discussion at a scientific meeting. One of my companions, an outstandingly successful Australian science journalist working mainly on radio, berated playwrights for the way in which they represented scientists as eccentric ineffective figures, not worthy of serious respect in the "real world." "That's funny," I said, "they do exactly the same to the clergy."

Science and theology face similar problems in relation to the public's perception of them. Yet they share one fundamental aim which will always make them worthy of the attention of those imbued with intellectual integrity and the desire to understand: in their different ways and in their different domains, each is concerned with the search for truth. In itself, that is sufficient to guarantee that there will continue to be a fruitful developing dialogue between them.

CHAPTER FIVE

Critical Realism
in Science and Religion

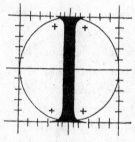

N the previous chapter I made the claim that physicists today know much more about the universe than did the most illustrious of their predecessors, Sir Isaac Newton. No immodesty was involved in that assertion; we are simply the benefactors of three hundred years of cumulative discovery. To the scientist no implausibility was involved in the assertion either, for surely the list of common knowledge at our disposal, but not at Sir Isaac's, is long and impressive: the details of the atomic structure of matter, the wave/particle duality of light, the molecular basis of chemistry and genetics, . . . One could go on and on. No doubt there are many topics of current concern on which the last word has not yet been spoken. Refinements are always possible and occasionally quite radical revision may be necessary in one aspect or another. Yet there is a robustness associated with well-tried concepts which discourages the expectation that they will disappear altogether. Atoms have surely come to stay! Although Maxwell's notion of the aether was discarded, his more basic understanding of electromagnetic waves is a persisting part of the modern theory of light. Maxwell's equations are still the equations of the electromagnetic field, though they are now understood in the sense of quantum field theory, with Maxwell's classical fields being simply a limiting case. Our maps of the physical world may reveal surprising new features as the scale and detail change, but they are superposable—we understand how the coarse-grained account given in Maxwell's classic *Treatise on Electricity and Magnetism* relates to the much more refined discussion of modern quantum electrodynamics. Newton himself further illustrates this point. His celebrated three laws of motion are not the basis for a modern understanding of fundamental (quantum) dynamics, but they are sufficiently verisimilitudinous when applied to large slowly moving bodies to provide totally adequate accounts for a great many purposes. Their limiting relationship with the equations of quantum mechanics is well understood through the so-called correspondence principle.

I am not making a triumphalist claim of majestic and untroubled scientific advance. Any realistic account will include setbacks as well as successes, revisions as well as reinforcements of current theories, puzzles as well as progress. I concluded a memoir I wrote on the basis of my experience of being a member of the high-energy physics community during the thirty-year period of the discovery of the quark structure of matter, by saying:

> In the short term it is a lurching story of both bafflement and breakthrough, with fluctuating fortunes, pseudoproblems, errors perpetrated, errors corrected, triumphant vindications and un-

expected discoveries. In the longer term, when there is time for an averaging out of experimental vagaries and a critical sifting of theoretical speculations, the character of the story is clearer. In 1950, the high-energy physics community believed protons and neutrons to be fundamental particles. In 1980, the high-energy physics community believed protons and neutrons to be composites composed of quarks and gluons. . . . In a word, the dust had settled.[1]

I then went on to defend this judgement in some detail. It is worth remarking that, although thirty years may seem a short time to cover in a study claiming some typicality as an instance of scientific method, the rate of advance of modern physics is such that 1950–80 represents a substantial episode in the history of discovery. The scientific generations crowd in on each other, and there is a clear sense in which we can say that most of modern science has taken place within this century. Those philosophers of science who rightly turn to the history of science for illumination need to remember this. It is not enough continually to revisit Galileo and Lavoisier.

Not all scientific proposals survive, but a well-winnowed domain of encounter with the physical world achieves a stable description and a reliable understanding. I think that Philip Kitcher puts it judiciously when he says, "The seeming growth in our understanding . . . is, I think, partly captured by the presence in later practices of an increasing number of stable reports of phenomena, with the rate of increase greatly outstripping the rate of revision."[2] He was referring to certain sciences relevant to evolutionary theory, but I think that the same assessment, with the addition of conceptual as well as

1. J. C. Polkinghorne, *Rochester Roundabout* (Longman/W. H. Freeman, 1989), 159 (somewhat condensed).

2. P. Kitcher, *The Advancement of Science* (Oxford University Press, 1993), 51.

phenomenal progress, is true much more widely. Like most scientists, I believe that the advance of science is concerned not just with our ability to manipulate the physical world, but with our capacity to gain knowledge of its actual nature. In a word, I am a realist. Of course, such knowledge is to a degree partial and corrigible. Our attainment is verisimilitude, not absolute truth. Our method is the creative interpretation of experience, not rigorous deduction from it. Thus, I am a critical realist.

The first order experience of the scientific community strongly encourages the sense of discovery, the belief that we are given to know more about the universe than was the privilege of our predecessors. In fact, without that belief, a great many of us would not have undertaken the long apprenticeship and weary labour which are an indispensable part of scientific research. Yet the second order reflections of the philosophers of science by no means unanimously endorse this judgement. When scientific method is put under the epistemological microscope, many feel that a flimsy structure is exposed. The difficulties are at least as old as David Hume's critique of the method of induction. How can general conclusions arise from particular experiences, however finitely many of these there may be? How can you be sure the sun will rise tomorrow? How can we gain real knowledge of what the physical world is actually like?

I have elsewhere reviewed, and responded to, some of the critiques of scientific realism made by twentieth-century philosophers.[3] I do not want to recapitulate that discussion

3. J. C. Polkinghorne, *Rochester*, chap. 21; *Beyond Science* (Cambridge University Press, 1996), chap. 2; see also J. Leplin, ed., *Scientific Realism* (University of

here, but I shall make some specific points drawn out from those considerations. The aim is to indicate the epistemological character of a commitment to critical realism in science.

First, we must resist "total account" theories of knowledge and be prepared instead to value more piecemeal achievements. By that I mean that we do not need to be right (or agree) about everything in order to be right (or agree) about some things. Revisions and precisions of detail should not be regarded as producing sharp discontinuities in understanding, where what is involved is clearly interpretable as a refinement of our knowledge of the nature of a common referent. J. J. Thomson pictured the electrons he discovered as being little hard lumps of matter, tiny charged "currants" in the atomic "pudding." I regard them as excitations in the quantum field of the electron. Yet we are clearly talking about the same entity (the light negatively charged constituent of atoms), and our descriptions differ only because the advance of knowledge has enabled me to share in a more exact (but not necessarily exhaustive) description. Those who resist this "charity of reference" will always be able to concoct too many stories of radical change, intending to use such tales to support claims of the irrationality of science. I regard this as a perverse and misleading strategy.

Second, it has proved impossible to distil the essence of the scientific method. Proposals such as making refutable conjectures (Popper), pursuing progressive research programmes (Lakatos), attaining empirical adequacy (van Fraassen) or pragmatic success (Rorty), capture aspects of the complex

California Press, 1984); W. H. Newton-Smith, *The Rationality of Science* (Routledge and Kegan Paul, 1981).

practice of science but each falls far short of an adequate account.[4] It has not proved possible to draw up a universal protocol for scientific research and I accept the verdict of Michael Polanyi (himself a successful scientist before he became a philosopher) that this is because science is an activity of persons, drawing on tacit skills learned through apprenticeship in a community whose purpose is the universal intent to seek truth about the physical world, while also acknowledging that current conclusions must remain open to the possibility of correction.[5] As part of this skilful practice, there is an essential role for the evaluation of non-empirical criteria such as economy, elegance, and naturalness, whose satisfaction is vital to the acceptance of a scientific theory. The justification for the use of these criteria lies in their having proved historically to be the means of identifying theories which turn out to have long-term fruitfulness, evidenced by their being capable of yielding understanding of phenomena not encountered or envisaged at the time of the theory's conception. The enforcement of these criteria proves, again as a matter of historical experience, to be the way in which the scientific community is able to solve the notorious problem of the underdetermination of theory by experiment. Rather than being faced with a plethora of possibilities, scientists find it a struggle to discover the one theory that proves acceptable on both empirical and non-empirical grounds.

Third, one must acknowledge that theory and experiment are inextricably intertwined in scientific thought, so that it is not possible to treat its advance in knowledge as re-

4. See note 3.
5. M. Polanyi, *Personal Knowledge* (Routledge and Kegan Paul, 1958).

sulting from the unproblematic confrontation of theoretical prediction with experimental fact. The operation of scientific instruments can be understood only in terms of scientific theory itself, a point at least as old as the controversy about whether Galileo's telescope was a reliable means of observing the lunar surface. Thus, there are no significant scientific facts that are not already interpreted facts. There is an inescapable self-sustaining circularity in the mutual relationship of theory and experiment, but I would claim that the understanding gained thereby proves that this circularity is benign and not vicious. The relationship between theoretical ideas and experimental measurements is not so plastic that the latter can be moulded to conform in their interpretation to theoretical notions chosen at whim. Rather, it is a delicate and difficult task to find mutually compatible theoretical and experimental accounts, so that when one is discovered it proves correspondingly convincing. The single great exception to this claim that scientists are led to an essentially unique interpretation of the consistent relationship between theory and experiment might seem to be provided by quantum theory. It is well known that the same empirical facts can either be understood in terms of an indeterministic theory (following Niels Bohr and his successors) or a deterministic theory (David Bohm).[6] This dilemma has been resolved by almost all physicists through an appeal to the non-empirical criteria discussed earlier, for they judge the Bohm theory to display a degree of unnatural contrivance when compared with the development of conventional quantum mechanics. It remains instructive, however, to

6. See, for example, the discussion of J. C. Polkinghorne, *Reason and Reality* (SPCK/Trinity Press International, 1991), chap. 7.

note that the controversy could not be settled on experimental grounds alone.

Fourth, the study of science encourages the recognition that there is no universal epistemology but rather entities are knowable only through ways that conform to their idiosyncratic nature. The simplest illustration of this fact is provided by quantum theory. Heisenberg's uncertainty principle forbids us the degree of clarity of knowledge to which we have access in the picturable world of classical physics. This does not make the quantum world unknowable, but it restricts us to a veiled encounter with quantum reality, met on its own terms.

Fifth, although social factors can accelerate or inhibit the growth of scientific knowledge (for example, by making some lines of enquiry either fashionable or unfashionable), they do not determine the character of that knowledge (once again I assert that the physical world is not so pliable that we can twist it into shapes that please our fancy). This claim would, of course, be contested by adherents of the "strong programme" in the sociology of knowledge. My reasons, nevertheless, for maintaining that it is true are ones which I have sought to give in the course of those earlier writings whose conclusions I am now summarising. They centre on the resistance of nature to our prior expectations and the consequent sense of discovery when order is found.

Sixth, and perhaps most important of all, the doctrine of scientific realism has been formulated as the best means of understanding our actual experience of doing science. For example, it affords the most natural way of understanding the sequence of investigations that led from atoms and molecules to quarks and gluons, if we suppose that these results should be conceived of as yielding a tightening grip on an existent reality. Similarly, the most natural explanation of science's impressive instrumental power to achieve technological ends is that it arises from science's actual knowledge of the nature of the entities it is manipulating. It is a contingent historical fact that our minds have proved apt to the discernment of nature's structure, with mathematics providing the key to that unfolding understanding. We can conceive of the possibility that the universe might have been opaque to our investigation, that our minds might have been too dull to grasp its inner coherence, or even that there might have been a world too disorderly to be coherent at all, a universe of magical whimsy in which the sun might or might not rise tomorrow. These are hypothetical possibilities which have been found not to be in accord with our actual experience. Scientific realism is a contingent fact about the relation between our epistemological powers and the ontology of our world, and not a metaphysical necessity concerning all possible worlds. There is no general inevitability about being able to make reliable and fruitful inductions on the basis of finite experience; it just happens to be the case that we are the kind of persons living in the kind of universe in which such logically precarious exercises are possible. A judicious philosophy of science is based on the analysis of particular contingent experience and not on the establishment of universal necessary truths. (Those who wish to understand more deeply this remarkable property of the rational transparency of our world should, in my opinion, first look to a theology of creation rather than to a philosophy of knowledge.) [7]

At the heart of scientific realism lies the conviction that

---

7. Cf. Polkinghorne, *Reason and Reality*, 76–77, and chap. 1.

intelligibility is the reliable guide to ontology, that concepts and entities whose postulation enables us to make deep sense of wide swathes of experience, are to be taken with the utmost seriousness as candidate descriptions of what is actually the case. While the resolute sceptic can never be defeated in logical argument, neither can the epistemologically optimistic who decline to despair of gaining verisimilitudinous knowledge of reality. It is the instinct of a scientist to encourage a trusting attitude towards those insights that afford a satisfying basis for understanding what is going on.

Are we to suppose that it is only in our investigations of the objective, impersonal physical world that we find ourselves endowed with these powers of apprehension, or may we be encouraged to trust our cognitive abilities across a much broader spectrum of human encounter with reality? As a passionate believer in the ultimate integrity and unity of all knowledge, I wish to extend my realist stance beyond science to encompass, among many other fields of enquiry theological reflection on our encounter with the divine. I take as my motto for that endeavour the remarkable words of Bernard Longergan: "God is the unrestricted act of understanding, the eternal rapture, glimpsed in every Archimedean cry of Eureka."[8] The search for truth through and through is ultimately the search for God. We need then to discuss how the six points made in relation to scientific realism have their counterparts in relation to theological realism.

First, we must consider charity of reference, the recognition that total agreement of description is not required before mutual discourse can take place and that the sharing of

8. B. Lonergan, *Insight* (Longman, 1958), 684.

knowledge and insight can initially be partial rather than total. Theologically, this raises the internal question of orthodoxy and heresy and the external question of how the world faith traditions relate to each other.

It is best to consider the internal question first. One cannot do so without repentance for the terrible and implacable actions of the past, mindful of Christians' recourse to war and to the persecution of those who were held to have deviated from the rule of faith. Today we are fortunate to live in more tolerant times, but that cannot mean that there are no limits to what may with honesty be contained within the description "Christian." While I myself believe that a coherent Christianity requires a strong doctrine of the incarnation, I can recognise that my colleagues who take a more functional view are seeking nevertheless to ascribe a pre-eminent significance to Jesus Christ and that this places them clearly within the Christian fold. Yet I cannot go so far as to think that non-realist accounts of God, which regard the divine as being simply an internalised symbol for individually chosen value, constitute an acceptable form of Christian belief. To say that is, of course, in no way to impugn the moral seriousness of those who take that view, nor to deny the importance that they assign to "religion" as they understand it. I simply do not think it is helpful to evacuate Christian theism of content to such an extent as to make the phrase too vague to be usefully discriminating.

The external question of inter-faith relationships arises in two parts. The Abrahamic religions—Judaism, Christianity, Islam—share a number of common features stemming from their interlaced histories. They are surely seeking to speak of the same God, even though they make many different assertions about the divine nature. Judaism lays revelatory emphasis on God's dealings with a chosen people. Christianity lays its revelatory emphasis on God's presence in Christ. Islam lays its revelatory emphasis on the infallible message of the Qur'an. Perplexing though this diversity is, it pales before the contrast with the religions of Eastern Asia, with many strands of Buddhism appearing not to hold a belief in a deity at all. Keith Ward describes the divide as being between Semitic religion (with its notions of God the Creator and of human beings living unique lives) and Indian religion (with its notions of moral entail [*karma*] and cosmic law [*dharma*] and of human beings undergoing a cycle of rebirths). In seeking some degree of reconciliation of these contrasting understandings, he suggests that "it seems that in almost every respect the Semitic and Indian traditions are complementary, emphasizing the active and unchanging poles respectively of the Supreme Spiritual Reality to which they both seek to relate."[9] I am not so easily persuaded that a deeply-lying compatibility can be discerned in this way or a synthesis achieved.[10]

Yet there are certainly reasons for supposing that, despite their many-tongued discourse, the world's religions are at least seeking to speak of a shared encounter with spiritual reality. Ward seeks to map out some of this common ground when he writes that "to have an insight into primary [that is, archaic] religion, and perhaps into any living religion, is to have some grasp of how one can have a form of awareness

9. K. Ward, *Religion and Revelation* (Oxford University Press, 1994), 331; see also K. Ward, *Images of Eternity* (Darton, Longman and Todd, 1987).

10. J. C. Polkinghorne, *Science and Christian Belief* (SPCK, 1994), published simultaneously as *The Faith of a Physicist* (Princeton University Press, 1994), chap. 10.

which is preconceptual, mediated largely through feeling and essentially imbued with value." [11]

Our appeal to charity of reference in science was largely in order to gain a diachronic understanding of how the claim of progress in knowledge was compatible with the eventual inadequacies of theories, resulting in the corrections and expansions that, in due course, they are found to require. In the case of theology one has to add to that the synchronic necessity to see how stable contemporary conflicting religious traditions can be understood to relate to each other. No doubt the vulnerability of deep personal experience to the historically formed varieties of cultural expression is part of the story of religious diversity. Another part is the necessary ineffability of the infinite God to finite human minds and the unavailability of the divine nature to being put to experimentally manipulated testing. Even after these considerations are taken into account, serious problems still remain that will require much more study and thought if progress is to be made with them. I have to acknowledge that the most severe challenge to theological realism comes from the conflicting variety of the world faith traditions. Yet my conviction that my Christian understanding affords me some access to the way things are, and to the One who is, means that I cannot relinquish that stance of critical theological realism, however great the perplexities which beset it in some respects.

Second, there is the recognition of the personal character of judgement that is involved in acquiring scientific knowledge, rendering the practice of science incapable of being

11. Ward, *Religion and Revelation*, 71.

encapsulated in a single methodological formula. This gives science a kinship with other forms of rational enquiry, for it is no longer perceived as possessing a unique means of access to reliable knowledge, unparalleled elsewhere. Its impressive success in answering questions to universal satisfaction is then seen to derive, not from the possession of utterly distinctive epistemological and ontological techniques, but from the comparative tractability of its subject material, an impersonal physical world open to repeated experimental manipulation, in contrast to the more subtle realms of unrepeatable experience which correspond to personal encounter and to the transpersonal meeting with the divine. The difference between science and its cousinly disciplines in the search for motivated belief is not of a fundamental kind but it lies in the degree of the power of empirical interrogation which these various investigations enjoy. The philosophical acknowledgement that there is no foundationally certain guarantee of scientific knowledge serves not to diminish the claims of science to verisimilitudinous success, but to encourage other modes of enquiry to comparable acts of intellectual daring in trusting the understandings that they attain by making sense of their experience.

Disciplines with restricted empirical access will have to depend more heavily on non-empirical factors in the assessment of the interpretations they propose. Coherence and comprehensiveness are factors of the highest importance in the search for metaphysical understanding. The fruitfulness to be appealed to in defence of such a strategy will, for theology, be more than simply the kind of development of doctrine discussed in chapter 2. It will also embrace the transformation of life. "You will know them by their fruits" (Mt. 7: 16). No one, mindful of the sad story of the wars of religion, will suppose that appeal to be unambiguous, but there is a real history of holiness within the history of the Church which provides a ground for treating its insights with spiritual seriousness. As an instance, consider the influences of the Rule of St. Benedict on Western European civilization over many centuries. The presence of a history of authentic spiritual experience of a most profound kind within each of the world faith traditions is the ground for supposing them to have important things to say to each other, despite the discord of their competing cognitive claims.

Third, the inescapable circularity of science, with theory and experiment, interpretation and event, in a mutually self-sustaining relationship, constitutes another form of cousinly connection with other rational disciplines. Theology has long known that one must believe in order to understand (commitment to a tradition is essential, for there is no neutral Archimedean point of detachment from which judgement can be made; insight is gained only through participation) and yet also one must understand in order to believe (faith is not the uncritical acceptance of dictated propositions; every spirit is to be tested [1 Thess. 5:21]; every image of God will in the end be found to be an inadequate idol). Once again, the success of science encourages the view that this circularity is benign, the means of knowledge and not the source of error, and so it emboldens other disciplines to essay a similar intellectual intrepidity.

Fourth, the recognition that there is no universal epistemology, and that our knowledge of entities must conform to their idiosyncratic natures, is an essential realisation in the framing of a just theology. God is to be known in ways that conform to the divine nature, different from the way in which any creature can be apprehended. This has been an important theme in the writings of Thomas Torrance: "We cannot begin by forming independently a theory of how God is knowable and then seek to test it out or indeed actualize it and fill it with material content. How God can be known must be determined from first to last by the way in which He actually is known." [12] There is a necessary obliqueness in the way in which the Infinite is met and known by finite beings.

In chapter 1, we considered some general aspects of human encounter with the world that encouraged the belief in a divine mind and purpose behind its history. This modest exercise in natural theology represents the use of reason and general experience in the search for God. In chapter 2, we considered the Christ-event and its aftermath. Here appeal was being made to a unique revelatory episode. The Anglican theological tradition in which I stand has always sought to add to the insights of reason and revelation, the further illumination obtained from tradition, which we can understand as deriving from the accumulated religious experience of the Christian community.

The identification and evaluation of religious experience has been the subject of much discussion. God is not just one entity among the many entities of the world, available to be picked out and examined in isolation. The divine presence is the ground of the world's being and the Creator is party

12. T. F. Torrance, *Theological Science* (Oxford University Press, 1969), 9.

to every occurrence. The infinite God is veiled from direct encounter with finite human beings. Hence the use of "mystery" as a word to emphasise the paradox of a meeting with divine ineffability. Such language is concerned with theological reality and not with obfuscation. Nicholas Lash draws an analogy with the lesser mystery of human personality: "Other people become, in this measure, 'mysterious,' not in so far as we *fail* to understand them, but rather in so far as, in lovingly relating to them, we succeed in doing so." [13]

The immanent omnipresence of God implies that there is not a limited, separable domain of experience which can be labelled "religious." John Hick wrote that "as ethical significance interpenetrates natural significance so religious significance interpenetrates both ethical and natural." [14] Another analogy one might make would be with the laws of nature, which operate everywhere and at all times. (Indeed I think they are pale expressions of the Creator's steadfast will.) There are, of course, particular events that we call experiments, in which the operation of these laws is most clearly manifested, just as there are focused moments of encounter with the divine (Elijah in the cave; Isaiah in the Temple; Peter, James, and John on the Mount of Transfiguration; Saul on the road to Damascus). The difference is that experiments can be contrived but moments of intense religious experience are given—we are back with Torrance's point about God's freedom of disclosure. Yet, just as we can be aware of natural laws outside the laboratory (meditating on the fall of an apple, for

13. N. Lash, *Easter in Ordinary* (SCM Press, 1988), 236.
14. J. Hick, *Faith and Knowledge*, 2d ed. (Cornell University Press, 1966), 112.

instance), so God is not confined to the realm of the conventionally sacred. Lash wrote his book to defend the thesis

> on the one hand, that it is not the case that all experience of
> God is necessarily religious in form or content, and on the other
> hand, that not everything which it would be appropriate to char-
> acterize, on psychological or sociological grounds, as 'religious'
> experience would thereby necessarily constitute experience of
> God.[15]

"The wind blows where it wills . . . so it is with everyone who is born of the Spirit" (John 3:8).

William James gave one of the most celebrated discussions of these issues in his Gifford Lectures, "The Varieties of Religious Experience." [16] He concentrated on the mystical experience of encounter and union with the One or the All, and his concern with saintliness made him pay particular attention to the religious geniuses, the pattern setters of the spiritual life. James repudiated an intellectualist approach to religion and his background as a distinguished psychologist encouraged him to focus on feeling as the essence of religious experience (following a tradition at least as old as Schleiermacher's concern with the feeling of absolute dependence). Lash criticises James's neglect of the conceptual, for experience is always interpreted, as the intertwining of theory and experiment in science illustrates. He also wishes to counterbalance James's concentration on the individual by recognising the role of the tradition of a community. (After all, Roman

15. Lash, *Easter*, 7.
16. W. James, *The Varieties of Religious Experience* (Collins, 1960); for an extended critique, see Lash, *Easter*, chaps. 2–8.

Catholics see visions of the Blessed Virgin Mary, Buddhists see visions of the Buddha.)

Although James's account is rather elitist in its concentration on the spiritual pattern setters, the surveys conducted by the distinguished biologist, Alister Hardy, show that significant numbers of "ordinary" people have had experiences of oneness with reality and of reassurance that all will be well. The settings for these events, and the people who participate in them, are often not what one would conventionally call "religious."[17] Richard Swinburne has argued that common principles of credulity, involving taking what witnesses say seriously, unless we have reasons to doubt their testimony, should encourage us to pay heed to this kind of experience.[18] No doubt particular care is called for in the area of religious experience—spiritual directors know how easily deception can creep in—but it is very difficult indeed to dismiss all such accounts.

I believe that the unitive experience of the mystic is an encounter with the divine in the mode of immanence. There is also the numinous encounter in the mode of transcendence—God met with in awe and reverence—described by Rudolf Otto in *The Idea of the Holy*.[19] For many ordinary believers, like the present author, religious experience will be in a lower key, mediated through sacrament, prayer, silence, and obedience. The liturgy offers a non-restrictive framework, expressing the

17. A. Hardy, *The Spiritual Nature of Man* (Oxford University Press, 1974).

18. R. Swinburne, *The Existence of God* (Oxford University Press, 1979), chap. 13.

19. R. Otto, *The Idea of the Holy* (Oxford University Press, 1923).

convictions of the traditional community but enabling a freedom of response within which the individual worshipper can find his or her particular place.

William Alston has framed his discussion of religious experience in terms of a defence of direct realism, the immediate awareness of the presence of God. He develops the notion of "doxastic practice" (forming beliefs and epistemically evaluating them) by first giving a careful analysis of sense perception.[20] A number of points emerge, some of them corresponding to issues we have already considered. There is an inescapable circularity within the doxastic practice, but this is capable of internal self-support through various kinds of fruitfulness. Alston concludes in general that "for any established doxastic practice it is rational to suppose that it is reliable and hence rational to suppose that its doxastic outputs are prima facie justified,"[21] a judgement delivered in the original with italic emphasis. He then goes on to consider religious "M-beliefs" (M for manifestation), being careful, however, to repudiate the "epistemic imperialism" which would fail to acknowledge the necessary individuality by which each doxastic practice conforms to its object material. Concerning Christian mystical practice he concludes that

> it possess a prima facie title to being rationally engaged in, and its outputs are thereby prima facie justified, *provided we have no sufficient reason to regard it as unreliable or otherwise disqualified for rational acceptance.*[22]

20. W. P. Alston, *Perceiving God* (Cornell University Press, 1991). There are some similarities with R. Swinburne (n. 18), but Alston emphasises the importance of social endorsement.

21. Ibid., 183.

22. Ibid., 225.

He finally goes on to consider possible signals of unreliability or sources of disqualification. The most troubling of these for Alston is provided by the diversity of world faith traditions. In the end, he believes that it is the life-enhancing aspect (fruits of the Spirit) of one's own religious tradition that completes the justification of one's adherence to it. To a realist, this individual note may seem a little too close to pragmatism for comfort.

Even so brief a survey shows that there is material on which critical realist theologians can base their study, though its assessment will need much greater subtlety and discrimination than are required in the analogous task faced by the scientist.

Fifth, one must acknowledge that social and cultural factors are clearly much more significant in their effects on religion than they are on science. The fact that religion is practised within a particular community formed by a particular tradition ensures that this will be so. I have already suggested that here is to be found part, but only a part, of the explanation for the diversity of the world faiths. I do not conclude, however, that this implies that theology is purely a social construction. I am encouraged in this belief by the insights of a natural theology that draws on the intelligibility and fruitfulness of the universe for its arguments and so looks outside the realm of existential human experience. Its relative cultural independence can be defended in a way similar to the defence of the relative cultural independence of the conclusions of science. Of course, such a natural theology falls far short in its deliverances of what would be needed for a fully articulated Christian theology, but it offers encouragement for the exploration of whether the divine might also be made known in the more ambiguous domains of personal encounter.

Sixth, there is the remarkable and fortunate fact that we are people of such an intellectual kind, living in a universe of such rationally transparent kind, that we are enabled to understand a great deal of the pattern and process of the world that we inhabit. Theologically, this is to be understood as due to the universe's being a creation and ourselves as creatures made in the image of the Creator. The possibility of science is then the consequence of the deposit of the imago dei within humanity. The critical realism which I have been seeking to defend is thus found, in one of those circularities which we have discovered to be inherent in the human search for knowledge, to be undergirded by a theological belief in the faithfulness of God, who has not created a world whose appearances will mislead the honest enquirer. The unity of knowledge is underwritten by the unity of the one true God; the veracity of well-motivated belief is underwritten by the reliability of God.

I believe that nuclear matter is made up of quarks which are not only unseen but which are also invisible in principle (because they are permanently confined within the protons and neutrons they constitute). The effects of these quarks can be perceived, but not the entities themselves. To borrow language from theology, we know the economic quark but not the immanent quark. Yet, on the basis of intelligibility as providing the grounds for ontological belief, a view which has already been defended in the scientific context, I am fully persuaded of the reality of the quark structure of matter. I believe that it makes sense of physical experience precisely because it corresponds to what is the case. A similar conviction grounds my belief in the invisible reality of God.

Colin Gunton has drawn our attention to a contemporary dilemma.[23] On the one hand is modernity's longing for foundationalism, the titanic quest for universal and certain knowledge—in Gunton's terminology, the search for the One. On the other hand is post-modernity's form of the assertion of anti-foundationalism, the dissolution of knowledge into private and particular points of view expressed through fideistic assertion or the playing of an idiosyncratic language game—in Gunton's terminology, the role of the Many. Today, the former does not convince, and the latter does not satisfy. With Polanyi as one of his guides, Gunton seeks a middle way. "The quest must therefore be for non-foundationalist foundations: to find the moments of truth in both of the contentions, namely that particularity and universality each have their place in a reasoned approach to the truth."[24] The end of that quest, I believe, is critical theological realism.

This idea of critical realism has been popular among those exploring the interaction between science and theology.[25] Wentzel van Huyssteen has given a careful general discussion of the issues, suggesting three criteria for a valid systematic-theological model of reality:

1. The reality depiction of theological statements.
2. The critical and problem-solving ability of theological statements.

23. C. Gunton, *The One, the Three and the Many* (Cambridge University Press, 1993), esp. chap. 5.

24. Ibid., 134.

25. See the references of chapter 2, n. 17. See also J. C. Polkinghorne, *Scientists as Theologians* (SPCK, 1996), chap. 2.

3. The constructive and progressive nature of theological statements.[26]

General meta-criteria have to be tested for their appropriateness against particular examples. Both the specific case of Christological enquiry in the first five centuries, discussed in chapter 2, and the somewhat more generalised comparisons of this chapter, lend their support to theology's being able to meet the challenge of these criteria.

Long ago, St. Anselm gave epigrammatic expression of critical realism in theology when he described theological enquiry as being *fides quaerens intellectum*, faith seeking understanding. Commitment to belief is the way to deepen and correct understanding; reality is made known to us as we trust the insights that come to us. We need not share Descartes' anxiety that we are being misled by a malicious demon, for the world is the creation of the God of truth who wills that our minds should have access, both to something of the divine nature and also to something of the wonders of the world that God holds in being. The scientist and the theologian both work by faith, a realist trust in the rational reliability of our understanding of experience.

26. W. van Huyssteen, *Theology and the Justification of Faith* (Eerdmans, 1989), 146.

CHAPTER SIX

Mathematical Postscript

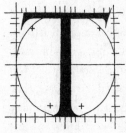

T HE argument of the first chapter was based on taking a generous and just view of the nature of reality, according as much significance to our subjective experiences of beauty and moral imperative as we do to our more objective encounters with the physical world. In chapter 3, I referred in passing to dual-aspect monism, the attempt to frame a metaphysical account which takes with equal seriousness the mental and the material poles of being, conceived as complementary phases of a single reality. I find that if I am in discussion with someone unwilling, at the least, to try to conceive of this wider view, then there is insufficient common ground for us to meet upon. "Dialogue" with an eliminative materialist, explicit or implicit in that belief, is just a talking past each other much of the time.

My main reason for defending this broad picture lies in what I perceive to be the self-authenticating character of the beautiful and the good which, while subjectively experienced, would seem grotesquely diminished and distorted if treated as mere epiphenomena. One must, of course, acknowledge the inescapably personal character of the assessment involved. While that does not worry me—I think our being persons is the most significant thing about us and something to be trusted—it clearly worries others. The purpose of this brief postscript is to explore another aspect of human experience, less vulnerable to individual tricks of perspective but equally indicative, I believe, of the need for a capacious understanding of reality. I refer to the human encounter with mathematical truths.

How are we to understand the character of mathematics? Is it just a form of mental gymnastics or is it an exploration of an already existent mental realm? Are mathematical truths invented or discovered? I believe that in the discussion of questions of this character we should give considerable weight to the testimony of those most closely involved. Of course, they are not necessarily infallible in their judgement, but they enjoy an immediacy of experience to which the second-order commentators should accord due respect. The mathematicians very frequently use the language of discovery. The internationally famous geometer Alain Connes writes, "Take prime numbers, for example, which, as far as I am concerned, constitute a more stable reality than the material reality that surrounds us. The working mathematician can be likened to an explorer who sets out to discover the world." [1] If one considers a profound, indeed inexhaustibly rich, mathematical entity such as the Mandelbrot set (frequently discussed in relation

---

1. J.-P. Changeux and A. Connes, *Conversations on Mind, Matter and Mathematics* (Princeton University Press, 1995), 12.

to fractals and chaos theory[2]), it is hard to believe that it came into being when Benoit Mandelbrot first began to consider its deceptively simple definition. As Roger Penrose says, "Like Everest, the Mandelbrot set is just *there*." He goes on to say, "There is something absolute and 'God given' about mathematical truth."[3]

Such views have not been uncontested, for there are a variety of philosophical assessments of the nature of mathematics. It has sometimes been thought of as an extension of logic, a grossly swollen construction of elaborate tautologies. This was the programme that Bertrand Russell and A. N. Whitehead sought to articulate in the desiccated pages of *Principia Mathematica*, and which was the motivation for David Hilbert's valiant attempt at the complete axiomatisation of mathematics. This heroic labour was shown to be abortive when the young Kurt Gödel proved that all axiomatic systems rich enough in structure to incorporate the natural numbers (the integers), contain stateable but undecidable propositions and that their self-consistency cannot be established. Mathematical truth is found to exceed the proving of theorems and to elude total capture in the confining meshes of any logical net.

Others have seen mathematics as being simply a human construction. This led Connes's conversation partner, the neurobiologist Jean-Pierre Changeux, to go so far as to make a physicalist response to the former's idealist view of mathematics by saying, "It seems to me, on the contrary, that mathe-

2. See J. Gleick, *Chaos* (Heinemann, 1988), 215–29.
3. R. Penrose, *The Emperor's New Mind* (Oxford University Press, 1989), 95, 112.

matical objects exist materially in your brain."[4] An austere version of the constructivist programme is the intuitionism of Luitzen Brouwer, who regarded the integers as directly accessible to the intuition of human reason and then required the rest of mathematics to be the result of finite constructive steps building on this foundation. He was following a programme set out in the nineteenth century by Leopold Kronecker who, notoriously, had said that the integers were made by God and the rest was the work of human beings.

Most mathematicians have found this to be far too narrow a concept of their subject. Kronecker's lifelong sparring partner, Georg Cantor (who discovered the beautiful theory of transfinite [infinite] numbers), took the Platonic view that mathematics is the exploration of an existing noetic realm. The Cambridge analyst G. H. Hardy spoke for many when he wrote,

> I will state my own position dogmatically . . . I believe that mathematical reality lies outside us, that our function is to discover or observe it, and that the theorems which we prove, and which we describe grandiloquently as our "creations," are simply our notes of our observations.[5]

Such an observational account accords intelligibly with the astonishing powers of mathematical intuition which were displayed by Hardy's protégé, the young Indian Ramanujan. He could write down deep mathematical results to which he had a kind of direct access which did not appear at all to involve calculated construction.

4. Changeux and Connes, *Conversations*, 13.
5. G. H. Hardy, *A Mathematician's Apology* (Cambridge University Press, 1967), 123–24.

One of the most uncompromising Platonists was Gödel himself. He once wrote concerning the mathematical theory of sets (a foundational subject which had been found to need very careful formulation if paradoxes were not to result),

> despite their remoteness from sense experience, we do have something like a perception also of the objects of set theory, as is seen from the fact that the axioms force themselves upon us as being true. I don't see any reason why we should have less confidence in this kind of perception, i.e. in mathematical intuition, than in sense perception, which induces us to build up physical theories and to expect that future sense perceptions will agree with them and, moreover, to believe that a question not decidable now has meaning and may be decided in the future. The set-theoretical paradoxes are hardly more troublesome for mathematics than deceptions of the senses are for physics.[6]

The arguments will continue, for deep metaphysical questions do not lend themselves to knock-down answering. There is a reminiscence here of the medieval debates between the realists and the nominalists. Nevertheless, I believe there is a much more persuasive case for believing in the reality of the Mandelbrot set than in the reality of the Idea of a lion. There is a realm of physical experience containing sticks and stones. There is also a realm of mental experience containing the truths of mathematics. These are not disjoint realms but they are parts of an interlinked complementary created reality, as our "amphibious" experience as embodied thinking reeds testifies, and as is also witnessed to by the "unreasonable effectiveness" of mathematical pattern as the clue to the structure of physical law (p. 4).[7] I believe that mathe-

6. Quoted in J. Barrow, *Pie in the Sky* (Oxford University Press, 1992), 261.
7. Cf. J. C. Polkinghorne, *Science and Creation* (SPCK, 1988), chap. 5.

matics provides a powerful—and for a scientist, readily accessible—encouragement to eschew physical reductionism and to embrace a generous view of the mental/material nature of reality.

Index

Active information, 62–63, 66–67, 93–94
Alston, W., 120–21
Anselm, St., 10, 124
Anthropic Principle, 6–9
Aquinas, St. Thomas, 10, 20, 58
Aspect, A., 28
Assimilation, 86
Augustine, St., 39, 69
Axiological argument, 20

Barbour, I. G., 40, 77–78, 86
Barth, K., 80
Benedict, St., 115
Biologists, attitudes of, 78–79
Block universe, 68–69
Boethius, 69
Bohm, D., 53, 67, 107
Bohr, N., 22, 107
Born, M., 27
Bottom-up thinking, 84
Brouwer, L., 128
Bultmann, R., 80

Cantor, G., 128
Chalcedonian definition, 30–31, 39–40

Chance and necessity, 5–6
Changeux, J.-P., 127
Chaos theory, xiii, 51–54, 60–62, 63–66, 96–97
Christology, 30–44, 86
Complexity theory, 96–97
Connes, A., 126–27
Consonance, 86
Councils, Church, 38
Critical realism, xiii, 29–30, 44–45, 52–53, 97–98, 101–24
Cross of Christ, 43–44
Cultural factors, 113, 121

Dante, A., 23
Darwin, C., 5, 77
Dawkins, R., 12, 14, 18, 77
Dennett, D., 77, 94–95
Descartes, R., 124
Design, 5
Determinism, 69–70
Dirac, P. A. M., 2, 27–28
Dissipative systems, 61
Divine action, xii–xiii, 48–49, 54–63, 71–75, 83
Divine knowledge, 73
Divine temporality, 69–71
Dual-aspect monism, 50, 125

Resurrection, 22–23, 33, 73
Rorty, R., 105
Rosen, N., 28
Russell, B., 127

Schrodinger, E., 27
Science and theology, inter-
    relationship of, xii, 29, 32–44,
    45–47, 86–88, 90–91, 93, 99–
    100, 123
Scientist-theologians, 84–85
Suffering, 14, 42–44, 79
Swinburne, R., 119

Theory and experiment, 106–7
Thomson, J. J., 105
Time, 67–69
Tipler, F., 21

Top-down causality, 58–59, 64
Torrance, T. F., 80–81, 116

Unity of knowledge, xiv, 24
Unpredictability, 50–52, 59

Value, 16–20, 23
van Fraassen, B., 105
van Huyssteen, W., 123
Vanstone, W. H., 74

Ward, K., 112
Watts, F., 80
Wave/particle duality, 26–27
Whitehead, A. N., 127
Wigner, E., 4
World faiths, 90–91, 111–13